T0261636

Convexity in the Theory of Lattice Gases

Princeton Series in Physics

edited by Arthur S. Wightman
and John J. Hopfield

Quantum Mechanics for Hamiltonians Defined as Quadratic Forms *by Barry Simon*

Lectures on Current Algebra and Its Applications
by Sam B. Treiman, Roman Jackiw, and David J. Gross

Physical Cosmology *by P. J. E. Peebles*

The Many-Worlds Interpretation of Quantum Mechanics
edited by B. S. DeWitt and N. Graham

The $P(\Phi)_2$ Euclidean (Quantum) Field Theory
by Barry Simon

Homogeneous Relativistic Cosmologies *by Michael P. Ryan, Jr., and Lawrence C. Shepley*

Studies in Mathematical Physics: Essays in Honor of Valentine Bargmann *edited by Elliott H. Lieb, B. Simon, and A. S. Wightman*

Convexity of the Theory of Lattice Gases
by Robert B. Israel

Convexity in the Theory of Lattice Gases

by

Robert B. Israel

With an Introduction by
Arthur S. Wightman

Princeton Series in Physics

Princeton University Press
Princeton, New Jersey 1979

Published by Princeton University Press, Princeton, New Jersey

In the United Kingdom:
Princeton University Press, Guildford, Surrey

Library of Congress Cataloging in Publication Data will be found on the last printed page of this book

Clothbound editions of Princeton University Press Books are printed on acid-free paper, and binding materials are chosen for strength and durability.

Printed in the United States of America
by Princeton University Press
Princeton, New Jersey

To Debbie

CONTENTS

INTRODUCTION

CONVEXITY AND THE NOTION OF EQUILIBRIUM STATE IN THERMODYNAMICS AND STATISTICAL MECHANICS

A. S. Wightman

The purpose of this introduction is to provide motivation and a historical setting for the problems treated by Israel in the following. The historical approach is not inappropriate because some of the questions discussed were also treated by the old masters of thermodynamics and statistical mechanics over the last hundred years; yet they are perennial, arising anew in the context of the statistical mechanics of each dynamical system. The modern theory acquires additional flavor when it is savored as part of this long development.

Here are some examples of such perennial questions.

a) What is the structure of the family of equilibrium states of a dynamical system?

b) To what extent is that structure restricted by the laws of thermodynamics?

c) How are metastable states related to equilibrium states?

d) Are there general mechanical procedures for finding the macroscopic observables that enable one to label equilibrium states?

This introduction sketches, in broad strokes, ideas and results bearing on such questions. It is hoped that by providing intuitively appealing pictures it supplements the precise mathematical theory of the main body of the text.

Thermodynamics in the Days Before there was any Systematic Statistical Mechanics

In his first two papers [1] [2] published in the Transactions of the Connecticut Academy of Sciences in 1873, J. W. Gibbs undertook to give a characterization of the family of equilibrium states of a physical system consistent with the laws of thermodynamics as definitively formulated by R. Clausius in 1850 [3]. For simplicity, Gibbs considered only the case of a so-called simple fluid, the adjective simple being essentially defined by the statement that the thermodynamic variables describing the state of a fixed mass of fluid at rest are

V	volume
p	pressure
U	energy
S	entropy

related by

$$TdS = dU + pdV . \tag{1}$$

Here T is the absolute temperature. Equation (1) is a partial expression of the first and second laws of thermodynamics.

Gibbs' distillation of the mixture of experiment and theory available to him can be expressed in current mathematical language as follows.

1) S and V, or alternatively U and V are coordinates for the manifold of equilibrium states of a simple fluid. (Gibbs' language here sounds amazingly like that used in the modern definition of continuous manifold. He says [4]: ... "Now if we associate a particular point in the plane with every separate state, of which the body is capable, in any continuous manner, so that states differing infinitely little are associated with points that are infinitely near to each other, the points associated with states of equal volume will form lines, which may be called lines of equal volume, ...".) V ranges over the interval $0 \leq V < \infty$ while S ranges either over $-\infty < S < \infty$ or some interval $S_0 \leq S < \infty$. Similarly, $-\infty < U < \infty$ or for

some U_0, $U_0 \le U < \infty$. In the following for simplicity we will treat the latter of each of these cases.

2) The energy, U, is a single-valued convex function of the entropy, S, and volume, V, for $S_0 \le S < \infty$ and $0 \le V < \infty$. U has first derivatives with respect to S and V. Similarly, the entropy, S, is a single-valued concave function of the energy, U, and volume, V, for $U_0 \le U < \infty$ and $0 \le V < \infty$. S has first derivatives with respect to U and V.

Explicitly, the convexity properties mean the following. U is *convex* as a function of S and V if for any $V_1, V_2 \ge 0$ and S_1 and S_2 satisfying $S_0 \le S_1, S_2 < \infty$, and any real number a, satisfying $0 \le a \le 1$:

$$U(aS_1 + (1-a)S_2, aV_1 + (1-a)V_2) \le aU(S_1, V_1) + (1-a)U(S_2, V_2) \tag{2}$$

S is *concave* as a function of U and V, if for any $V_1, V_2 \ge 0$, and U_1 and U_2 satisfying $U_0 \le U_1 < \infty$ and $U_0 \le U_2 < \infty$, and any real number, a, satisfying $0 \le a \le 1$

$$S(aU_1 + (1-a)U_2, aV_1 + (1-a)V_2) \ge aS(U_1, V_1) + (1-a)S(U_2, V_2) \,. \tag{3}$$

3) U is, for fixed V, a monotonically increasing function of S so that the absolute temperature computed from equation (1)

$$T = \left(\frac{\partial U}{\partial S}\right)_V \tag{4}$$

is positive. Equation (1) also yields the connection between the pressure, p, and the slope of the U function in the direction of V axis

$$p = -\left(\frac{\partial U}{\partial V}\right)_S \,. \tag{5}$$

Alternatively, these statements may be expressed in terms of the entropy function, S,

$$\frac{1}{T} = \left(\frac{\partial S}{\partial U}\right)_V \ge 0 \tag{6}$$

$$\frac{p}{T} = \left(\frac{\partial S}{\partial V}\right)_U . \tag{7}$$

These statements 1), 2), 3) call for an extended series of comments.

Gibbs seems to have been the first to recognize the distinction between the information conveyed by the functions

$$(S,V) \to U(S,V) \qquad (U,V) \to S(U,V)$$

and that conveyed by the customary *equation of state* which is a relation between the pressure, volume, and temperature of equilibrium states of the form

$$f(p,V,T) = 0 \tag{8}$$

for some function, f. He called the former functions *fundamental equations* [4] (a notation that has not been widely accepted) because from them all thermodynamic properties of the fluid can be deduced by derivation. For example, since equation (4) gives T as a function of V and S we can compute from it

$$\left(\frac{\partial T}{\partial S}\right)_V (S,V) = \left(\frac{\partial^2 U}{\partial S^2}\right)_V (S,V) \tag{9}$$

and this in turn yields an expression for the heat capacity at constant volume

$$C_V = \left(\frac{dQ}{dT}\right)_V = T\left(\frac{\partial S}{\partial T}\right)_V = \frac{T}{\left(\frac{\partial T}{\partial S}\right)_V}$$

i.e., (10)

$$C_V = \frac{T}{\left(\frac{\partial^2 U}{\partial S^2}\right)_V} .$$

On the other hand, the equation of state in the variables p,V,T contains no such information as is evident from the example of an ideal gas. Its equation of state is

$$pV = NkT \tag{11}$$

where N is the number of moles of gas and k is Boltzmann's constant

$$k = 1.380 \times 10^{-23} \text{ joule}/{}^{0}\text{K} . \tag{12}$$

The heat capacity of the ideal gas is

$$C_V = aN \tag{13}$$

where a is an arbitrary constant not fixed by the equation of state.

As Gibbs explained [5] the convexity of the energy function, U, and the concavity of the entropy function, S, are consequences of the stability requirements imposed on equilibrium states by the second law of thermodynamics. What is equally extraordinary is that a converse statement holds. Any substance whose energy function, U, satisfies the requirements 1), 2), 3) is a simple fluid whose behavior is compatible with the second law of thermodynamics. No wonder that Maxwell was so enthusiastic about Gibbs' first two papers that he wrote a whole new section dealing with their results for the later editions of his book *Theory of Heat* [6].

For those who like their physics stated in simple general mathematical terms the version of thermodynamics offered by Gibbs' first two papers can scarcely be improved. Nevertheless, apart from its impact on Maxwell, it had very little influence on late nineteenth century textbooks. The notion of "fundamental equation" and the simple expression it gives for the laws of thermodynamics in terms of convexity and monotonicity only became available with the publication of "neo-Gibbsian" textbooks and monographs in the midtwentieth century [7] [8]. The point is also illustrated by the history of the construction of S, U, V surfaces for instructional purposes. Almost every physics laboratory has a model p, V, T surface for water. The first S, U, V surface was constructed by Maxwell who sent it as a gift to Gibbs [9]. As late as the 1930's such surfaces were still regarded as sufficiently novel that descriptions of their construction were published [10] [11]. Even in the volume devoted to a critical evaluation of Gibbs' work [12], the first two papers [1] [2] on thermodynamics are dismissed with the words "As papers I and II are properly characterized by H. A. Bumstead in his introductory biography as of importance not so much for any place they made for themselves in the literature as for the preparation and viewpoint they afforded the author as groundwork for his great memoir on *The Equilibrium of Heterogeneous Substances*, it will be most appropriate to illustrate them by an outline of Gibbs' course on thermodynamics as he gave it towards the end of his life." However enlightening and useful the following summary of Gibbs' course may be, it does not provide an adequate evaluation of the simple but profound message of papers I and II: *the equilibrium states of a simple thermodynamic substance form a once differentiable manifold which is the graph of a distinguished convex function, the energy,* U, *as a function of the entropy,* S *and the volume* V.

Since the role of convexity is so important here, it is worth sketching how it arises in the laws of thermodynamics [13]. For this purpose it is convenient to adopt the point of view of Gibbs' third paper [14] (which is really a book in itself) and indicate explicitly the total quantity of fluid under consideration, N. It will be measured in moles. Then the entropy in an equilibrium state is a function of U, V, and $N, (U,V,N) \to S(U,V,N)$, which is homogeneous of first degree

$$S(\lambda U, \lambda V, \lambda N) = \lambda S(U,V,N) \tag{14}$$

for any $\lambda > 0$, and $U_0 \leq U < \infty$, $0 \leq V < \infty$, $N \geq 0$, because all the quantities under discussion S,U,V,N are extensive. This assumption is part of the laws of thermodynamics.

Suppose one is given two isolated systems, 1 and 2, each in equilibrium, whose entropy, energy, volume, and mole number are aS_1, aU_1, aV_1, aN_1 and $(1-a)S_2$, $(1-a)U_2$, $(1-a)V_2$, $(1-a)N_2$ respectively. Here a is some real number such that $0 \leq a \leq 1$. If the two systems are brought into contact with a wall (very penetrable!) between them which admits the flow of heat as well as changes of volume and mole number but with conservation of total energy, volume and mole number, they will come to a final unique equilibrium state $S_{12}, U_{12}, V_{12}, N_{12}$ where

$$\begin{aligned} U_{12} &= aU_1 + (1-a)U_2 \\ V_{12} &= aV_1 + (1-a)V_2 \\ N_{12} &= aN_1 + (1-a)N_2 \end{aligned} \tag{15}$$

and, according to the laws of thermodynamics (in "Neo-Gibbsian" form [13])

$$S_{12} = \sup_{\substack{U', U'', V', V'', N', N'' \\ U'+U'' = U_{12} \\ V'+V'' = V_{12} \\ N'+N'' = N_{12}}} [S_1(U', V', N') + S_2(U'', V'', N'')] \tag{16}$$

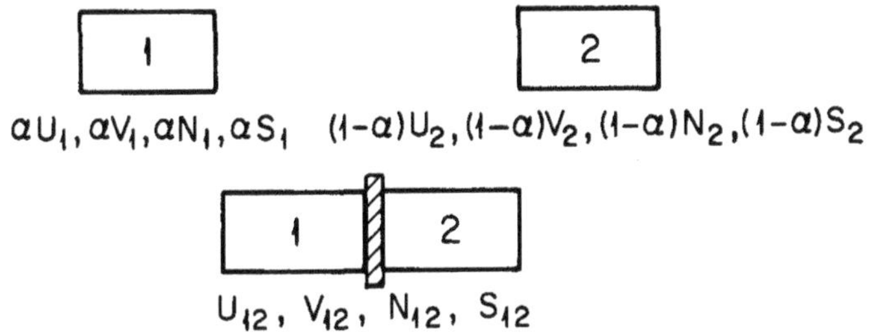

Fig. 1. Configurations for the experiment displaying the physical meaning of the concavity of the entropy function.

This inequality clearly implies

$$\begin{aligned} &S_{12}(aU_1 + (1-a)U_2, aV_1 + (1-a)V_2, aN_1 + (1-a)N_2) \\ &\quad \geq aS_1(U_1, V_1, N_1) + (1-a)S_2(U_2, V_2, N_2) . \end{aligned} \tag{17}$$

If the two systems are of the same material and we take for granted that as a result of the approach to equilibrium we get a homogeneous phase, then the functions S_1, S_2, and S_{12} are all the same, say S, and this inequality becomes the defining relation for the concavity of S. Conversely, if S is any concave function of U, V, N then a standard theorem assures us it is continuous and a straightforward argument, which uses that S is assumed homogeneous of first degree, yields

$$S(U, V, N) = \sup_{\substack{U', U'', V', V'', N', N'' \\ U'+U''=U \\ V'+V''=V \\ N'+N''=N}} [S(U', V', N') + S(U'', V'', N'')] \qquad (18)$$

Thus, the concavity of S implies this relation, typical of those which express the consequences of the laws of thermodynamics for S. The argument for the rest of these relations is similar, so the concavity of S is not only necessary but also sufficient in order that the laws of thermodynamics hold for this fluid. There is an analogous argument in which the convexity of U as a function of S, V, N replaces the concavity of S as a function of U, V, N.

In his second paper Gibbs explored the consequences of convexity for the coexistence of phases of simple fluids. He recognized that convex bodies (like the set of points S, U, V space lying above the graph of U or those points lying below the graph of S) have boundaries with a very special structure. A point, P, of the boundary is either an *extreme point* in which case there is no open line segment passing through P consisting entirely of points of the boundary or it is an interior point of a *face*. (See Figures 2 and 3.) The structure of faces is, in general, complicated but there are simple cases which have an elegant physical interpretation.

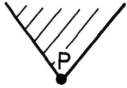

Fig. 2. Examples of extreme points in the boundary of convex bodies. For the body indicated by the striped regions at the left, P and every other boundary point are extreme points. For the body at the right, P is the only extreme point.

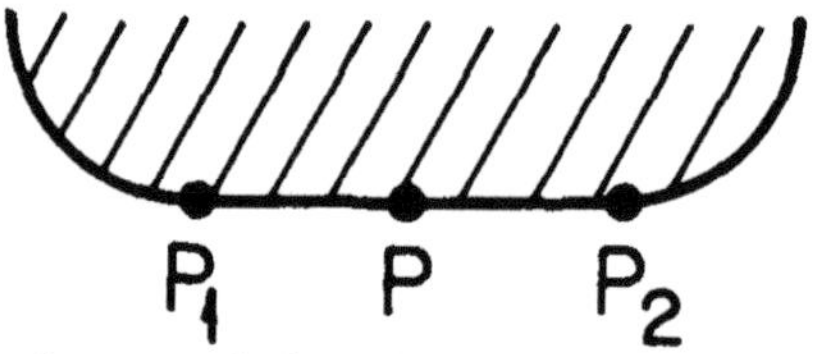

Fig. 3. P lies in the face consisting of the closed line segment connecting P_1, P_2. The convex body is the striped region.

Suppose $P_1 = \{S_1, U_1, V_1\}$ and $P_2 = \{S_2, U_2, V_2\}$ are extreme points and the line segment P_1P_2 joining them lies in the boundary. (See Figure 3.) Then if $P = \{S,U,V\}$ is the point of the line segment a fraction a of the way from P_1 to P_2

$$\begin{aligned} S &= aS_1 + (1-a)S_2 \\ U &= aU_1 + (1-a)U_2 \qquad 0 < a < 1 \\ V &= aV_1 + (1-a)V_2 \, . \end{aligned} \tag{19}$$

Thus, $\{S,U,V\}$ are the parameters of a state deserving the name *mixture* of the coexistent phases $\{S_1, U_1, V_1\}$ and $\{S_2, U_2, V_2\}$. On the other hand, the $\{S,U,V\}$ of an extreme point permits no such decomposition and deserves the name *pure phase*. Note that the fact that all mixtures obtained by varying a in (19) have the same pressure and temperature is a consequence of the formulae (4) and (5) and the fact that the tangent plane at P_1 osculates along the whole segment P_1P_2. This geometrical interpretation of coexisting phases of simple fluids extends to *triple points* at which the mixtures are formed from three coexisting pure phases $P_i = \{S_i, U_i, V_i\}$, $i = 1,2,3$ (say gas, liquid, and solid). Then the boundary contains as a face the triangle whose vertices are P_1, P_2, P_3 and any mixture is of the form

$$\{S,U,V\} = \sum_{i=1}^{3} a_i \{S_i, U_i, V_i\} \tag{20}$$

with

$$a_i \geq 0, \quad i = 1,2,3 \, ; \quad \sum_{i=1}^{3} a_i = 1 \, .$$

Gibbs also pointed out that the notion of critical point has a neat geometrical interpretation. The boundary may contain a piece of ruled surface, which itself contains no two-dimensional planar subset as shown in Figure 4.

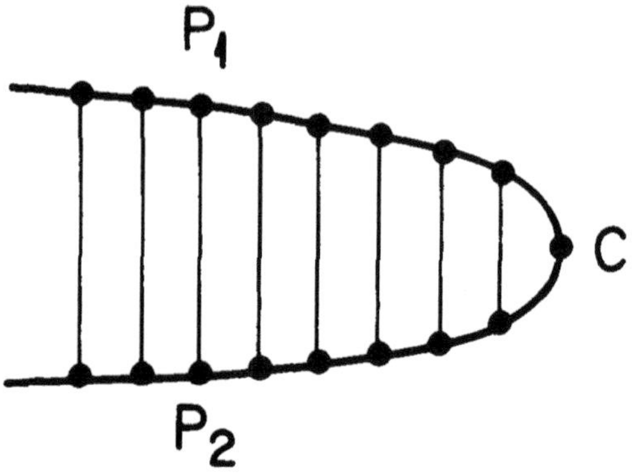

Fig. 4. A piece of ruled surface lying in the boundary of a convex body. P_1 and P_2 are coexisting pure phases, i.e., they are extreme points of the boundary but the line connecting them lies in the boundary. C is a critical point in the sense that each neighborhood of it contains a pair of coexisting distinct pure phases. The figure is intended to show how the ruled part of the surface looks as seen from inside the body.

The line segments are faces whose end points, P_1 and P_2 for example, are extreme points. C is a *critical point* (in the sense of thermodynamics) because it is an extreme point in every neighborhood of which there exists a pair of distinct coexisting pure phases. The geometric possibilities of triple points and coexistence regions of pure phases are combined in the typical configuration associated with the liquid gas and solid phases of a simple fluid, shown in Figure 5.

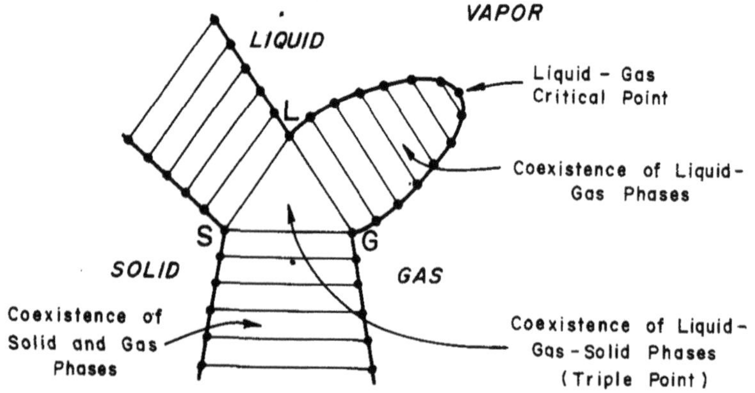

Fig. 5. A typical configuration of points in $\{S,U,V\}$ space describing the phases of a simple fluid which has liquid, gas, and solid phases. The indicated triangle and pieces of ruled surface are flat places in the graph of the convex function $U : \{S,V\} \to U(V,S)$. The diagram is intended to show how the surface defined by this function looks when seen from above.

Gibbs also took a position on the description of *metastable* states. He presumed that, when a fluid possesses metastable states, the surface formed by the extreme points has a continuation into the coexistence region that is distinct from the ruled surfaces describing coexisting pure phases and that this extended surface is locally but not globally convex. Gibbs argued that this latter property implied stability against continuous phase changes but instability against the discontinuous change into a mixture of liquid and gas phases.

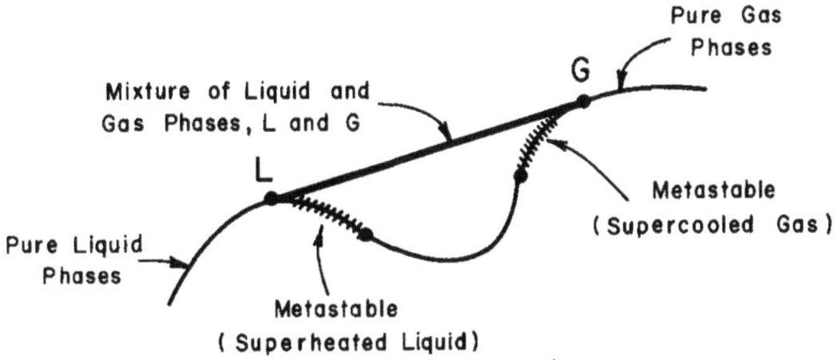

Fig. 6. Slice through the entropy surface of a simple fluid in the coexistence region of liquid and gas showing how the surface of metastable phases extends the surface of pure phases.

Gibbs' interpretation of metastability is one of the few results in his work on thermodynamics which is in doubt as a result of more recent work. It is currently believed that there may be manifolds of equilibrium states which have coexistence regions like that shown in Figure 6, but the surface of pure liquid and pure gas phases may have no natural continuation beyond L or G, respectively. The matter is still under study [15].

If one asks to what extent Gibbs proved that any of these phenomena follow from thermodynamic first principles, the answer is: not at all. What Gibbs did was to show how a variety of geometric possibilities occurring in the theory of convex bodies have a natural physical interpretation. He arranged the discussion of the principles of thermodynamics so the basic convexity properties came out easily. In the light of the century of physics that has elapsed since Gibbs wrote his three papers on thermodynamics, one can see how extraordinary his instinct for the "correct" conceptual framework really was [16].

The third Gibbs paper on thermodynamics is a wide-ranging discussion of chemical and physical phenomena ranging from the theory of capillarity and electromotive forces to catalysis and critical phenomena. For our purposes here, we single out a very few topics important for the later discussion of statistical mechanics.

In the third paper, Gibbs dealt with a thermodynamic system with an arbitrary finite number n of component substances so the internal energy of the system was generalized to be a convex and once-differentiable function U of n+2 real variables $S, V, N_1 \cdots N_n$ where $S_0 \leq S < \infty$, $0 \leq V < \infty$, $0 \leq N_1, \cdots, N_n < \infty$. He introduced and used systematically the chemical potentials

$$\mu_j = \left(\frac{\partial U}{\partial N_j}\right)_{S,V,N_1 \cdots \widehat{N_j} \cdots N_n} . \qquad (21)$$

μ_j is a measure of the rate of increase of the internal energy of the system as the number of moles, N_j, of the jth component is increased.

Pure phases are defined (just as for a simple thermodynamic substance) as extreme points of the convex surface defined by the function U. Again the criterion for the coexistence of two pure phases labeled by I and II is that there be a line segment lying in the surface whose endpoints are I and II. One has then, evidently,

$$p_I = p_{II}; \quad T_I = T_{II}; \quad \mu_{jI} = \mu_{jII}, \quad j = 1 \cdots n \qquad (22)$$

and indeed constant pressure, temperature, and chemical potentials for all points on the line segment. Gibbs' criterion for equilibrium in terms of the chemical potential gives a particularly neat derivation of the Maxwell rule for the isothermals of a liquid vapor system (see Figure 7).

The geometric possibilities for flat parts of the graph of a convex function of more than two variables are very rich and so one expects a corresponding richness in the possible configurations of coexisting phases of which we will illustrate a few to make clear how the convexity and differentiability of U leads directly to a natural history classification of phase transitions.

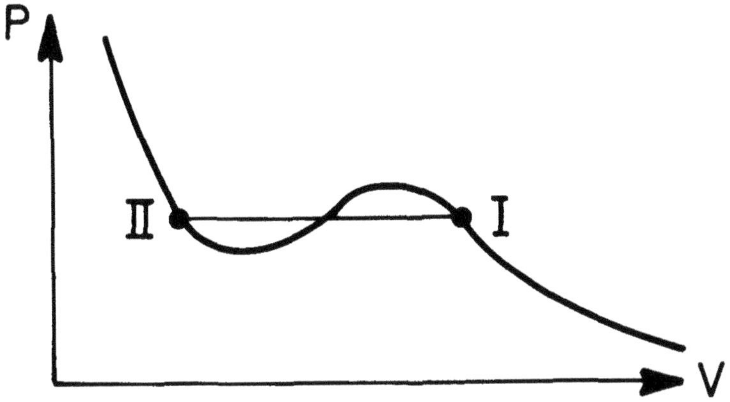

Fig. 7. *Maxwell's Equal Area Rule.* The isothermal has a flat portion. The criterion of equilibrium (22) says that if I and II are coexisting pure phases $\mu_I = \mu_{II}$, $p_I = p_{II}$, $T_I = T_{II}$. But

$$0 = SdT - Vdp + Nd\mu .$$

Thus integrating from I to II along the curved isothermal with N fixed, one has

$$0 = -\int_I^{II} Vdp + N(\mu_{II} - \mu_I)$$

so the area above the straight line between I and II is equal to the area below. (The differential relation follows from the homogeneity of U and the first law of thermodynamics, (1). Differentiating $U(\lambda S, \lambda V, \lambda N) = \lambda U(S,V,N)$ with respect to λ and setting $\lambda = 1$, one has $U = TS - pV + \mu N$ (the Gibbs-Duhem relation). Subtracting $dU = TdS - pdV + \mu dN$ from the differential of this relation yields the required result.)

Generalizing the discussion of triple points in Figure 5, we note that the simplex of largest dimension that can lie in the graph of the function, U, for a system with n components is $n+1$. Such a simplex describes an *(n+2)-ple point*. For a two-component system, this is a tetrahedron; it is illustrated in Figures 8 and 9. From the edges $A'B'$ and $A''B''$ etc., there issue ruled surfaces formed of lines $D'E'$, $D''E''$ etc., describing two coexisting phases. The $D'D''$ pure phases and the $E'E''$ pure phases form two surfaces separated by a line of critical points $F'F''$ etc. If we assume similar phenomena for the edges $A'C'$, $A''C'', \cdots$ and $B'C'$, $B''C'', \cdots$ then tC_1 is a *tricritical point*, the intersection of three distinct lines of critical points. Figure 9 shows the full configuration around a

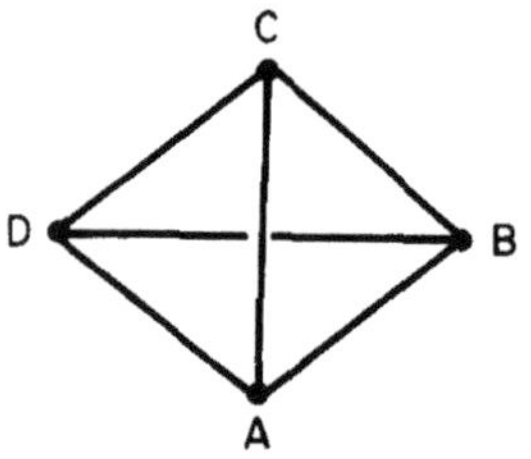

Quadruple point; four coexisting pure phases represented by the corners of the tetrahedron. The surface of u as a function of s,v,x_1 is flat over the tetrahedron.

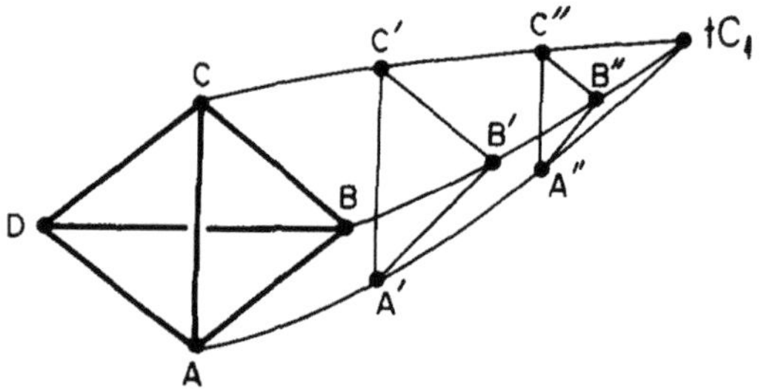

From the triangular face ABC, there issues a family of triangles $A'B'C'$, $A''B''C''$ etc., representing triple points; the vertices of each triangle are coexisting pure phases. The surface of U is flat over each triangle but varies from one triangle to another.

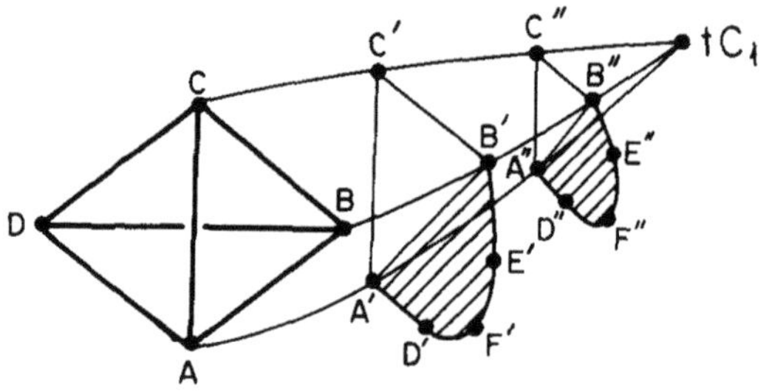

Fig. 8. Configurations of coexisting phases for a two-component system in s,v,x_1 space, in the neighborhood of a quadruple point. Here we have used the reduced variables $u = U/N$, $s = S/N$, $v = V/N$, $x_1 = N_1/N$ where $N = N_1 + N_2$.

quadruple point. It is the three dimensional analogue of the configuration around a triple point on a two-dimensional surface as shown on the left hand side of Figure 11. Of course, in practice, it may happen, just as it did in the one component case, Figure 5, that some of these tricritical points lie at infinity; even the quadruple point itself may lie there.

Gibbs drew somewhat different figures to describe the phenomena illustrated in Figures 8 and 9, because he introduced systematically the Legendre transforms of U with respect to various variables. For completeness, we digress to define them here in modern notation.

The *enthalpy* is

$$H = U + pV \tag{23}$$

which, because

$$dH = TdS + Vdp + \sum_{j=1}^{n} \mu_j \, dN_j ,$$

he regarded as a function of $S, p, N_1, \cdots N_n$.

The *Helmholtz free energy* is

$$F = U - ST \tag{24}$$

which, because

$$dF = -SdT - pdV + \sum_{j=1}^{n} \mu_j \, dN_j ,$$

he regarded as a function of $T, V, N_1, \cdots N_n$.

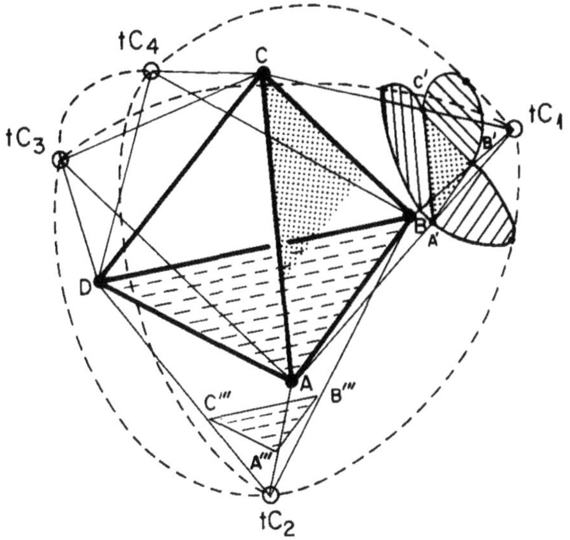

Fig. 9. The full configuration around a quadruple point. ABCD is the tetrahedron representing the quadruple point. From each of its four faces there protrudes a stack of triangles representing triple points, two such triangles are shown, $A'B'C'$ and $A'''B'''C'''$. Each stack of triangles ends in a tricritical point; these are labeled tC_1, tC_2, tC_3, tC_4. Each tricritical point is the intersection of three critical lines; these are the dashed lines. A two dimensional slice through the configuration is also shown, it has been chosen to include the triangle $A'B'C'$ and two phase coexistence lines. It is similar to the left hand side of Figure 10.

The *Gibbs free energy* is

$$G = U - ST + pV \tag{25}$$

which because

$$dG = -SdT + Vdp + \sum_{j=1}^{n} \mu_j \, dN_j$$

he regarded as a function of $T, p, N_1, \cdots N_n$.

$-p$ can also be regarded as a Legendre transform, but of U/V rather than U. If U/V is regarded as a function of the variables $S/V, N_1/V, \cdots N_n/V$, and $-p$ as a function of the variables $T, \mu_1, \cdots, \mu_n$ [17], we have

$$-p = \frac{U}{V} - T\frac{S}{V} - \mu_1 \frac{N_1}{V} - \cdots - \mu_n \frac{N_n}{V}. \tag{26}$$

Here there arises a small pedagogical point which is traditionally passed over by text-book writers. Do (23)...(26) actually define the enthalpy, H, as a function of $S, p, N_1, \cdots, N_n$, the Helmholtz free energy, F, as a function of $T, V, N_1, \cdots, N_n$ the Gibbs free energy, G, as a function of $T, p, N_1, \cdots, N_n$ and the pressure, p, as a function of $T, \mu_1, \cdots, \mu_n$? On the face of it, to compute H, for example, one is instructed to solve

$$p = -\left(\frac{\partial U}{\partial V}\right)_{S, N_1, \cdots N_n} (S, V, N_1 \cdots N_n) \tag{27}$$

for V as a function of $S, p, N_1, \cdots, N_n$ and to insert it in (23). This, in fact, is legitimate if U is *strictly convex*, i.e., if strict inequality holds in the defining relation (2) of convexity for $0 < \alpha < 1$. But what is to be done at a flat place in the graph of U? There, as the arguments of p vary, p remains constant, so (27) surely cannot be solved for V. The answer is that there is a definition of $H, F, G, -p$ which works for all convex U and reduces to the definition just discussed when U is differentiable and strictly convex:

$$H(S,p,N_1 \cdots N_n) = \inf_{V} [U(S,V,N_1,\cdots,N_n) + pV] \tag{28}$$

$$F(T,V,N_1 \cdots N_n) = \inf_{S} [U(S,V,N_1,\cdots,N_n) - TS] \tag{29}$$

$$G(T,p,N_1 \cdots N_n) = \inf_{S,V} [U(S,V,N_1,\cdots,N_n) + pV - TS] \tag{30}$$

$$-p(T,\mu_1,\cdots,\mu_n) = \inf_{S/V,N_1/V,\cdots,N_n/V} [\tfrac{1}{V}(U(S,V,N_1,\cdots,N_n) - TS - \mu_1 N_1 - \cdots - \mu_n N_n)] \,. \tag{31}$$

Here inf [ ] means the greatest lower bound of the quantity in brackets as the indicated variables run over their domains.

Starting from these definitions we can recover the formulae (23)... (26) at any point where U is strictly convex. At such a point, there is a unique value of the argument where the square bracket reaches is minimum. It can be determined by differentiating the bracket with respect to the variables and setting the derivatives equal to zero. That yields for H, for example, precisely equation (27).

What happens at the flat places in the graph of U may be seen by studying the derivatives of the new functions with respect to their arguments assuming such derivatives exist. For example, for p as a function of $T,\mu_1,\cdots,\mu_n$ one has

$$\left(\frac{\partial p}{\partial T}\right)_{\mu_1 \cdots \mu_n} = \frac{S}{V}\,; \qquad \left(\frac{\partial p}{\partial \mu_j}\right)_{T,\mu_1 \cdots \widehat{\mu_j} \cdots \mu_n} = \frac{N_j}{V}\,. \tag{32}$$

As one approaches the flat places in the graph of U the right hand sides have, in general, different limits for distinct coexisting pure phases. In order for there to be a difference in entropy of two coexisting pure phases $\left(\frac{\partial p}{\partial T}\right)_{\mu_1 \cdots \mu_n}$ must have a discontinuity; in order for there to be a difference in the density of the jth component, $\left(\frac{\partial p}{\partial \mu_j}\right)_{T,\mu_1 \cdots \widehat{\mu_j} \cdots \mu_n}$ must have a discontinuity. One has the geometrical situation illustrated on the right hand side of Figure 2: there is a corner or edge in the graph of p as a function of $T, \mu_1 \cdots \mu_n$.

Further light is thrown on the geometrical character of these discontinuities if one introduces a typical nineteenth century geometrical notion: the duality between points and hyperplanes relative to a conic section. Since such matters are rarely discussed in modern physics, I give a little detail illustrating the ideas for points and lines in the plane.

Let $\{x^*, y^*\}$ be any point of the plane. Then the *dual of* $\{x^*, y^*\}$ *relative to the unit circle*

$$x^2 + y^2 = 1 \tag{33}$$

is defined to be the straight line

$$xx^* + yy^* = 1 . \tag{34}$$

If $\{x^*, y^*\}$ lies outside the unit circle the dual has a simple geometrical construction: there are exactly two lines through $\{x^*, y^*\}$ tangent to the unit circle, the dual is the line through their points of tangency. See Figure 10.

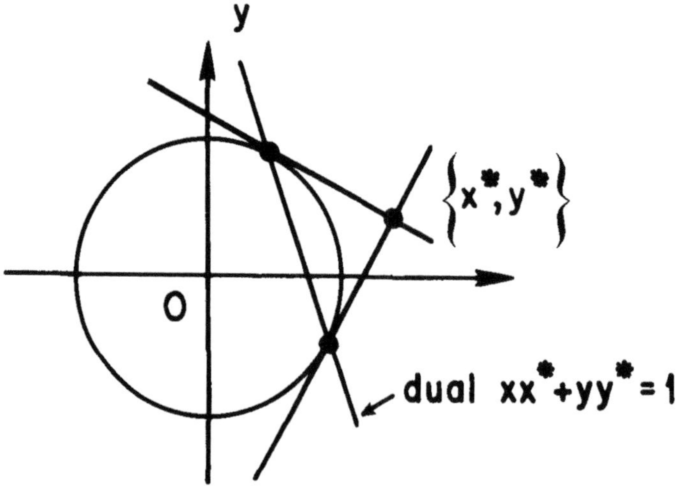

Fig. 10. Construction of the dual line, $xx^*+yy^* = 1$, of the point $\{x^*, y^*\}$ relative to the unit circle $x^2+y^2 = 1$.

If one makes an arbitrary projective transformation of the plane:

$$x \to \frac{a_1 x+b_1 y+c_1}{a_3 x+b_3 y+c_3} \qquad y \to \frac{a_2 x+b_2 y+c_2}{a_3 x+b_3 y+c_3}$$

one can transform the unit circle into an arbitrary conic section

$$Ax^2 + By^2 + 2Cxy + 2Dx + 2Ey + F = 0 . \tag{35}$$

Applying the same projective transformation to $\{x^*, y^*\}$, one sees that the dual line goes into

$$Axx^* + Byy^* + C(xy^* + x^*y) + D(x+x^*) + E(y+y^*) + F = 0 \tag{36}$$

which as x and y vary, with x^*, y^* fixed, is again a straight line. It is, by definition, *the dual of the point* $\{x^*, y^*\}$ *relative to the conic section* (35). Since projective transformations preserve tangency and carry straight lines into lines, the analogue of the construction given in Figure 10, works here also.

Now let us apply this definition for the particular conic section

$$x^2 - 2y = 0 . \tag{37}$$

It is easy to choose the projective transformation to obtain it. Take

$$\vec{a} = (a_1, a_2, a_3) = (1,0,0)$$

$$\vec{b} = (b_1, b_2, b_3) = \frac{1}{\sqrt{2}} (0, +1, 1)$$

$$\vec{c} = (c_1, c_2, c_3) = \frac{1}{\sqrt{2}} (0, -1, 1) .$$

Then the equation of the dual is

$$xx^* - (y+y^*) = 0 . \tag{38}$$

As we will now see this duality yields a helpful geometric insight into the relation between a convex function, f, of the variable, x, and its *conjugate convex function*, f^*, defined by

$$f^*(x^*) = \sup_x [xx^* - f(x)] .$$

(To anticipate, the functions H, F, G, $-p$ are just the negatives of such conjugate convex functions in one or more variables.) We study the sets lying above the graphs of the functions, f, and f^* (sometimes called their *epigraphs*). It is easy to see that the convexity of a function is equivalent to the convexity of its epigraph. Thus, to verify the convexity of f^*, we have only to verify the convexity of its epigraph. We will construct it as an intersection of half-spaces, its convexity follows immediately since intersection of convex sets is convex. Consider a point $\{x,y\}$ of the epigraph of f; its dual is the line in the $\{x^*, y^*\}$ plane that has as epigraph the half space

$$y^* \geq xx^* - y .$$

If y approaches f(x) from above the corresponding half spaces are decreasing with parallel boundaries. Thus, their intersection as y varies

$$\bigcap_{y \geq f(x)} \{\{x^*, y^*\};\ y^* \geq xx^* - y\}$$

is

$$\{\{x^*, y^*\};\ y^* \geq xx^* - f(x)\} .$$

Thus the intersection of the half spaces associated by duality with points of the form $\{x,y\}$ with $y \geq f(x)$ is just the half space associated with the single point $\{x, f(x)\}$. Finally, the intersection of these sets as x varies is

$$\bigcap_x \{\{x^*, y^*\};\ y^* \geq xx^* - f(x)\} = \{\{x^*, y^*\};\ y^* \geq \sup_x \{xx^* - f(x)\}\} .$$

The last being precisely the epigraph of f^*. *The graph of* f^* *is the envelope of the family of hyperplanes*

$$y^* = xx^* - f(x)$$

as $\{x, f(x)\}$ *varies over the graph of* f. If

$$\{x^{(\alpha)}, f(x^{(\alpha)})\} = \alpha\{x^{(1)}, f(x^{(1)})\} + (1-\alpha)\{x^{(0)}, f(x^{(0)})\} \qquad 0 \leq \alpha \leq 1 \tag{39}$$

are the points of a line segment lying in the graph of f, then

$$y^* = x^{(\alpha)}x^* - f(x^{(\alpha)}) \tag{40}$$

is a family of lines passing through a single point $\{x^*, f^*(x^*)\}$ of the graph of f. (The required x^* is such that the term proportional to α in (40) has zero coefficient.) Here is the promised refined geometric interpretation of what happens at a flat place in the graph of f: *as* $\{x, f(x)\}$ *varies on a straight line, the dual lines rotate around a corner in the graph of* f^*, *the convex combination* (39) *corresponding to the convex combination*

$$y^* = [\alpha x^{(1)} + (1-\alpha)x^{(0)}]x^* - [\alpha f(x^{(1)}) + (1-\alpha)f(x^{(0)})] \,. \tag{41}$$

The relationship between a convex function and its conjugate convex function, described here for one dimension, easily generalizes to convex functions on $\mathbf{R}^n$. Amazingly enough it did not appear in the mathematical literature as a precise mathematical theorem until 1948! [18]. It is stated in the next section. The theorem applies immediately to our functions H, F, G, and $-p$ as defined in (28)...(31); they are all concave. Gibbs was as ambiguous as all the other authors of that day (and most since), in his use of the "definitions" (23)...(26). However, given these "definitions," he proved the concavity and established (28)...(31) which then appear as minimum principles.

The same geometric possibilities for coexistent phases of a one-component system in terms of the energy U as a function of volume V and entropy S for fixed number of moles N are shown in terms of U/V, S/V, N/V and p, T, μ in Figure 11. Flat places in the graph of U/V go over by duality into points in the graph of p at which the tangent plane is not unique. For the quadruple point configuration of Figure 9, the dual is given in Figure 12.

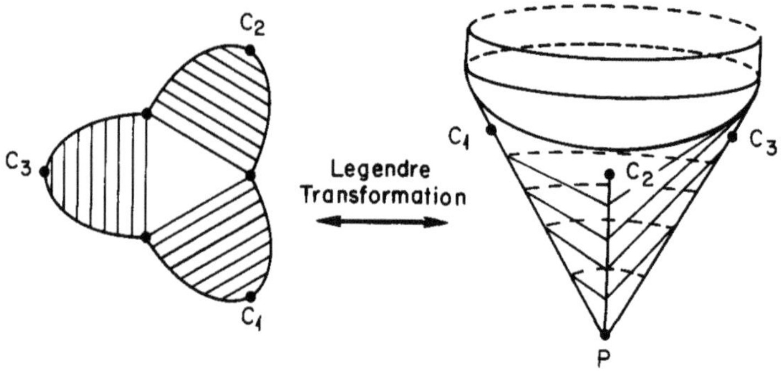

Fig. 11. Description of coexisting phases of a simple thermodynamic substance in terms of $\{U/V, S/V, N/V\}$ and $\{p, T, \mu\}$ variables. The triangle (triple point) on the left corresponds to the point P on the right. The figure on the right has been drawn to display the three edges meeting at the triple point P. The slopes involved are actually restricted by the inequalities $\frac{\partial p}{\partial \mu} = \rho \geq 0$ for all tangent planes. $\rho \geq 0$ follows from our initial assumption $N \geq 0$, $V \geq 0$. The three pieces of ruled surface on the left correspond to the three edges PC_1, PC_2, PC_3 on the right. The critical points C_1, C_2, C_3 on the left correspond to the points C_1, C_2, C_3 on the right.

Gibbs gave a description of a manifold of coexisting phases in a system with n components, the so-called Gibbs phase rule. He said "A system of r coexistent phases, each of which has the same n independently variable components is capable of $n+2-r$ variations of phase" [19]. His argument for this uses the $p, T, \mu_1 \cdots \mu_n$ diagram. He notes that the manifold for a single phase has dimension $n+1$ and the coexistence of two phases is defined by a hypersurface. To insure the coexistence of r phases requires that the phase lie on the intersection of $r-1$ such hypersurfaces. This, in general, yields a manifold of dimension $n+1-(r-1)$.

This argument as presented is not an airtight deduction, because $r-1$ hypersurfaces may intersect in a manifold of less or more than $n+1-(r-1)$ dimensions. From this point of view, Gibbs' result should be stated: Gibbs' Phase Rule holds except in those special cases where it is false! Gibbs recognized this because he said, later on, "... if $r = n+2$, no variation in the phases (remaining coexistent) is possible. It does not seem probable that r can ever exceed $n+2$."

A precise discussion of the Gibbs Phase Rule raises fundamental issues of thermodynamics. Should one regard the laws as laid down by Clausius as complete or should one adjoin other principles? For example, Figure 13 shows an internal energy function of a simple fluid which possesses a continuous infinity of coexisting pure phases. We have already remarked that the simplex of maximal dimension lying in the u surface is a triangle. Is there anything in the laws of thermodynamics which

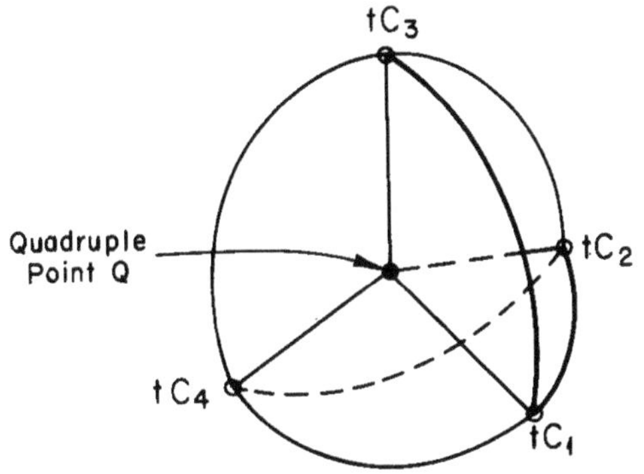

Fig. 12. The configuration in $\{p, T, \mu_1, \mu_2\}$ space arising from the Legendre transform of the configuration of Figure 9 in u, s, v, x_1 space. The projection on $\{T, \mu_1, \mu_2\}$ is shown. Each of the points tC_1, tC_2, tC_3, tC_4 is a tricritical point. Each is the endpoint of a line of triple points $QtC_1 \cdots QtC_4$ respectively, as well as an intersection point of three curves of critical points $tC_j, tC_k, j \neq k$. Each of the lines of triple points QtC_j, $j = 1, \cdots, 4$ is the intersection of three surfaces each of which is a two-phase coexistence surface whose edge is a curve of critical points. It is worth noting that even though only three planes are in general required to define a point in three dimensions, the quadruple point is the intersection of six surfaces.

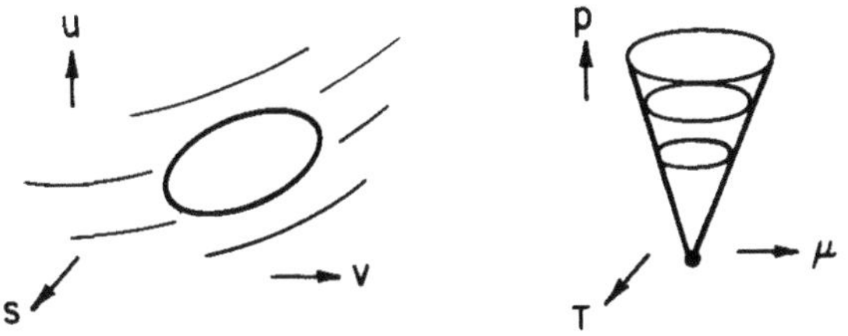

Fig. 13. The u-surface of a simple fluid violating *Gibbs' Phase Rule*. The circular disc lying in the u-surface describes a continuous infinity of coexisting phases. Its Legendre transform is shown on the right.

says that a square or circular disc or some other convex two-dimensional region should not be a maximal plane subset of the surface? The answer is no. The laws of thermodynamics, as usually stated, only require that u be a convex once-differentiable function of s and v. Now, it sounds physically plausible that a mixed phase should have a unique decomposition into pure phases. If one admits this principle as an addition to the laws of thermodynamics then such violations of the Gibbs phase rule as that displayed in Figure 13 are excluded by definition. Perhaps, more generally, one should simply adjoin the Gibbs Phase Rule to the laws of thermodynamics? However, proposals of this kind are almost certainly too strong to be reasonable. The simplest way to see this is to consider a composite system obtained from two non-interacting copies of a system which has a critical point at $T = T_c$ [20]. If the system has an internal energy $\{S,V\} \to U(S,V)$, then the corresponding free energy is

$$F(T,V) = \inf_S [U(S,V) - TS] .$$

We assume that the coexisting pure states, say $\{T, V'(T)\}$ and $\{T, V''(T)\}$ for $T < T_c$ have the same free energy. (That is unusual for simple fluids but common for the analogous coexisting phases in magnetic systems discussed in the next section.)

The free energy of the composite system is, by definition

$$F(T, V_1, V_2) = A(T, V_1) + A(T, V_2) . \tag{42}$$

Then the composite system has, for $T < T_c$, four coexisting pure states as one sees from Figure 14. These states lie on the corners of a square in the graph of the free energy for the composite system and thus violate the Gibbs Phase Rule.

Gibbs Phase Rule is reinterpreted in the context of statistical mechanics in quite a different manner. One considers a family of microscopic theories parametrized by their Hamiltonians and for each derives a set of thermodynamic functions. The statistical mechanical Gibbs Phase Rule

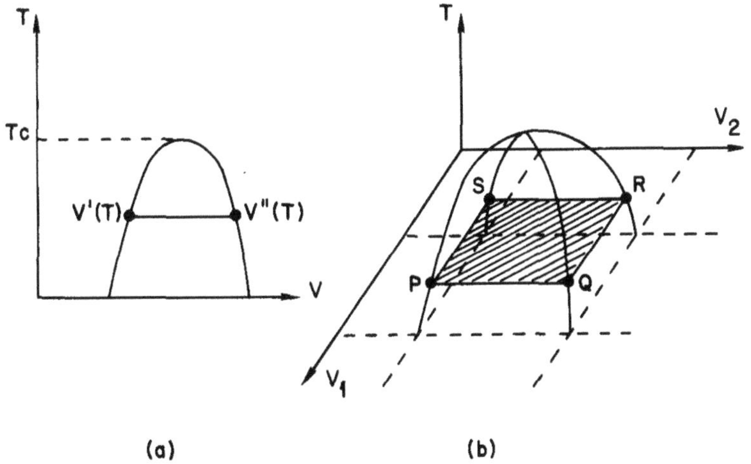

Fig. 14. a) Phase diagram of a hypothetical simple fluid.
b) Phase diagram of the composite system formed from two copies of a).

then says, roughly, that those Hamiltonians for which the thermodynamic functions fail to satisfy Gibbs' original formulation are a sub-family which is very small in some appropriate sense. (The adverb "roughly" appears here because in statistical mechanics the notion of equilibrium state is refined to include correlation functions and consequently this statement has to be refined.) This gives a precise meaning to the "improbability" of a violation of the phase rule but a meaning which depends in principle on the family of Hamiltonians considered. *Each family of microscopic dynamics has its own thermodynamics.* As will be discussed later, for some families of dynamics mixed phases always form a simplex and so always possess a unique decomposition into coexisting pure phases. An apparent contradiction with model of Figure 14 is avoided because the notion of equilibrium state is changed. See remarks b) under *Refinement of the Notion of State ...* below.

The phase diagrams discussed by Gibbs are quite general, since they indicate general possibilities constrained only by the laws of thermodynamics. Contemporaneously with Gibbs, quite a different approach was taken by Van der Waals [21]. He derived an equation of state describing a liquid-gas phase transition, which, when supplemented by the Maxwell equal area rule (described in Figure 7) gives a quantitative model in semi-quantitative agreement with experiment. The parameters of the model are related to the volume of the molecules of the fluid and their mean free path. A striking feature of Van der Waals' equation of state is its universality: when the pressure, temperature, and volume are expressed in units of the critical pressure, temperature and volume all parameters disappear from the equation. The year after its publication by Van der Waals, Onnes realized that this *law of corresponding states* is much more general than Van der Waals' equation itself [22]. In fact, it turns out to be a general prediction of classical statistical mechanics for particles interacting with a potential that has a simple transformation law under scale transformations. In any case, it has been a fundamental phenomenological tool to this day.

Van der Waals extended his theory to mixtures by permitting the basic parameters of his equation to depend on the mole fractions N_j/N. His theory then gave an expression for the Helmholtz free energy which, like that for a simple fluid, had to be supplemented by a Maxwell Rule. This Van der Waals did, following the principles laid down by Gibbs and seen in Figure 6: he took the convex envelope of his Helmholtz free energy function to get an equation of state. He and various of his co-workers showed that with appropriate choice of parameters one could, in fact, produce some of the geometrical possibilities for coexisting phases which Gibbs had shown are compatible with the laws of thermodynamics.

Later Developments in Thermodynamics

The preceding section dealt primarily with the work of Gibbs and Van der Waals. Later on, the manifold of equilibrium states was studied for a wide variety of other systems among which we single out three as of special interest for our purposes: magnets, $^3He - {}^4He$ mixtures, and mixtures of three or more liquids.

The simplest model of a magnet assumes that the internal energy is a function of the magnetization M with conjugate thermodynamic variable, the magnetic field H [23]. Thus

$$dU = TdS - pdV + HdM . \tag{43}$$

As is well known, all the phenomena associated with a phase transition can be beautifully illustrated on the ferromagnetic phase transition associated with the variables H and M. However, in the case of antiferromagnets, the bulk magnetization vanishes, and it is the total magnetization of a *sub*lattice which actually has to be identified with M. Then the conjugate magnetic field is a so-called staggered field which takes different values on the sublattice and its complement. The main point for our present purposes is not the detailed definition of these quantities but the fact that to understand the thermodynamics of an anti-ferromagnet, in particular, the singularities of its free energy as a function of temperature, it is necessary to know the free energy as a function of variables which are not really macroscopic. (The sublattice magnetization is indirectly observable in the scattering of neutrons by the lattice, but it is difficult to see how to produce a staggered magnetic field.)

The situation is even more striking for $^3He - {}^4He$ mixtures. There it is known that in addition to the phase transition arising from the superfluidity of ^4He there is at $T \sim 0.86\ {}^0K$ a further phase transition into two liquid phases, one rich in ^3He, the other poor. The customary phase diagram in T and $x = N_3/(N_3+N_4)$ is shown in Figure 15 along with the corresponding diagram in T and $\mu_3 - \mu_4$ where μ_3 is the chemical potential of ^3He and μ_4 that of ^4He [24]. Now the λ line is a line of critical points but as it stands, the diagram gives little indication that tC should be called a tricritical point. If, however, a "superfluid order parameter," ψ, and its thermodynamic conjugate, ζ, are introduced, then simple theoretical models indicate, that the free energy, F, of the fluid mixture as a function of T, $\mu_3 - \mu_4$, and ζ has a phase diagram indicated in Figure 16. The point tC then appears as an endpoint of a line of triple points as well as an endpoint of three lines of critical points. Through the line of triple points there pass three planes of coexistent

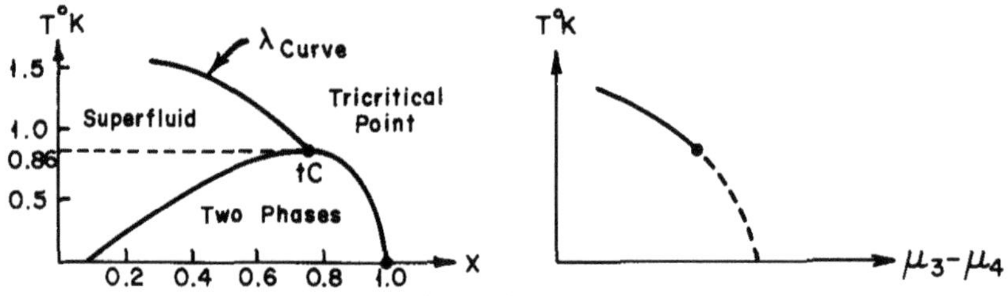

Fig. 15. The phase diagram of a ^{3}He – ^{4}He mixture, in the T–x plane and T – $(\mu_3-\mu_4)$ planes. Here x is the mole fraction $N_3/(N_3+N_4)$ while μ_3 and μ_4 are the chemical potentials of ^{3}He or ^{4}He respectively.

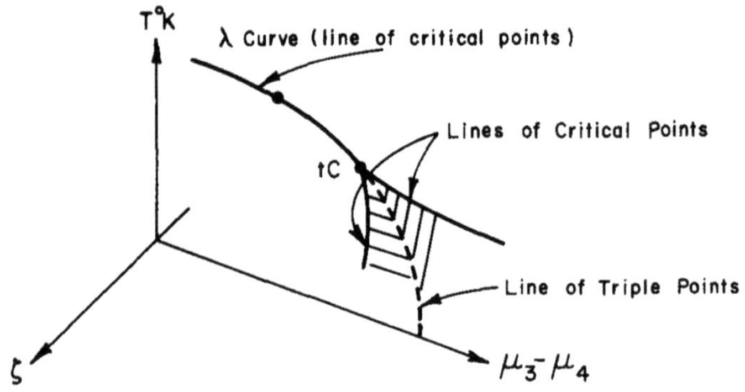

Fig. 16. The phase diagram of a ^{3}He – ^{4}He mixture in T, $\mu_3-\mu_4$, ζ space. Just as in Figure 10 the tricritical point appears as the endpoint of three lines of critical points and a line of triple points. ζ is the thermodynamic variable conjugate to the superfluid.

phases, one of the real world with $\zeta = 0$ and coexisting superfluid ^{4}He and ordinary fluid of ^{3}He and ^{4}He, the others purely theoretical with $\zeta > 0$ and $\zeta < 0$ respectively and coexisting ^{3}He rich and ^{3}He poor phases.

Thus, one has a natural extension of the phase diagram to include the dependence on the purely theoretical variable ζ.

The lesson to be learned from these two examples is that to obtain a unified description of phase diagrams one must expect to deal with parameters which are not, strictly speaking, macroscopic observables. The theory of tricritical points in fluid mixtures was developed in analogy with that which worked in ^{3}He – ^{4}He mixtures [25] [26] [27]. It is a striking fact that the geometric configuration used is quite different from that which leads to the tricritical points in Figure 9 and 16. The first step is to describe the notion of *critical end point.* It is a pure phase that lies on a line of critical points for the coincidence of two phases as well on a line of triple points describing coexistence of the two phases with a third. Such a configuration is shown in Figure 17. A critical end point in a system of two components is a common occurrence, but, according to the

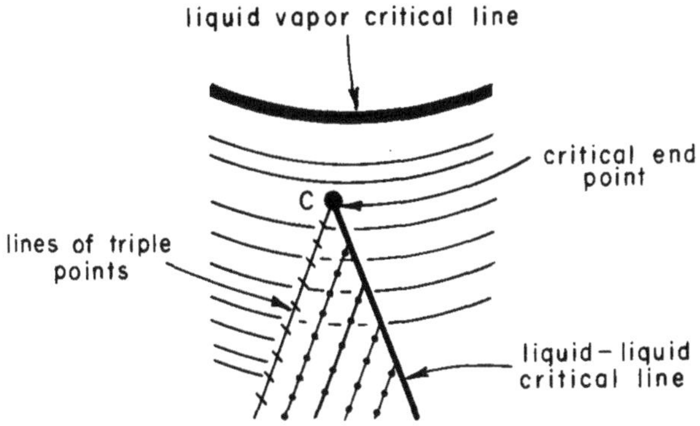

Fig. 17. A critical end point in a space of field variables, say T, μ_1, μ_2, for a binary system which two liquid phases and a vapor phase.

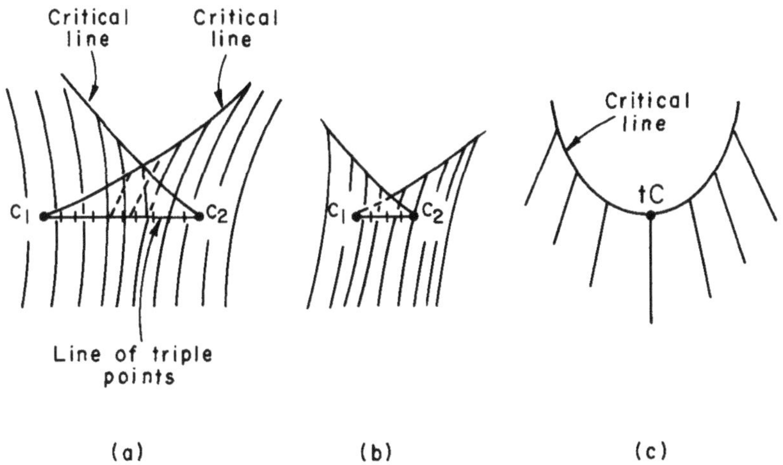

Fig. 18. A tricritical point, tC, in a three component liquid mixture. Configurations in $\{T, \mu_1, \mu_2\}$ space for μ_3 are shown: a) far from the critical value b) close to the critical value c) at the critical value.

notions of probability used in these models, for the critical endpoint, C, of Figure 17 to lie on the liquid vapor critical line would be a "lucky accident"; the point C would then be a tricritical point. For this to happen "non-accidentally" one needs at least one more variable μ_3. Then, as μ_3 varies, C will in general move until it collides with the critical line. The configuration that Griffiths actually found has an additional property; as μ_3 varies and C approaches the tricritical point, the line of triple points gets shorter and shorter. See Figure 18. The question of what constitutes a "lucky accident" may not be well defined unless one has specified the family of admissible thermodynamic functions. For example consider a simple fluid and suppose it is known that it has a triple point and that all differentiable convex functions are admissible. Then the two phase coexistence regions shown in Figure 11 have probability zero because they require the constancy of derivatives on certain families of lines and functions that satisfy such conditions are surely infinitely less numerous by any reasonable measure than those whose derivatives have only to satisfy convexity inequalities. This is not a serious argument but it does show by reductio ad absurdum that a specification of admissible thermodynamic functions is necessary before the notion of "lucky accident" is well defined. The question of the correct extension of the Gibbs Phase Rule to cover critical points deserves a study in the context of statistical mechanics.

It is very nice to have an extended phase diagram for $^3He - {}^3He$ mixtures like Figure 16 that brings out the analogy with Figure 12, but the real test of its usefulness comes when one examines the behavior of thermodynamic functions in the neighborhood of the tricritical point. That brings us to the next of the newer developments in thermodynamics: non-classical critical behavior.

It was already evident in the first decade of this century that while Van der Waals' equation gave a striking semi-quantitative account of phase transitions in gases and liquid mixtures, it failed in its account of critical behavior. For example, it gives for the coexistence curve in the neighborhood of the critical density $p - p_c \sim |\rho - \rho_c|^{\beta}$ with $\beta = 1/2$ instead of what is actually observed about 1/3 [28]. There is a large collection of critical exponents associated with the critical behavior of various other thermodynamic functions and almost all of them show departures from the classical or Van der Waals' values. Furthermore, the last decade and a half of experimental work has given growing evidence that the critical exponents have some universality properties, being apparently independent of the particular details of interaction and dependent only on the dimension of space and symmetry of the system. To explain these universality properties is obviously a fundamental problem of statistical mechanics. What a good thermodynamic description can do is to provide convenient parametrizations of the thermodynamic functions which is compatible with the laws of thermodynamics. Since, as we will see, statistical mechanics leads automatically to extended phase diagrams of the type discussed above, we have yet another reason for enlarging the scope of discussion of phase diagrams.

There is one last point about the manifold of equilibrium states that deserves some comment to complete the discussion of thermodynamics. That has to do with smoothness. We already know that when phase transitions occur the Van der Waals' prediction for the manifold of equilibrium states is only piecewise smooth. In fact, it is piecewise analytic. Is there any reason to believe that such smoothness or analyticity properties should be general features of the manifold? Indeed, to begin with, is there any reason to believe that this question ever occurred to the founders of thermodynamics? The answer to the latter question appears to be no. In fact, an appreciation of the difference between an infinitely differentiable and an analytic function was not widespread in mathematical physics until the second half of the twentieth century, after the advent of L. Schwartz's theory of distributions. One gets the feeling, reading Gibbs, that he is implicitly assuming piecewise analyticity but the only evidence I can adduce for this is his discussion of metastability in which he assumes that the manifold of equilibrium states continues uniquely into the region of metastability. That is weak evidence because the extension involved need not be given by analytic continuation.

As to the answer to the first question, from the perspective of today, it appears to be yes under appropriate conditions, and no under other appropriate conditions. Models in statistical mechanics can be made to display both smooth and non-smooth behavior as will be explained later. What thermodynamics requires, as has been emphasized, is only convexity plus the existence of one derivative of the internal energy, U, as a function of $S, V, N_1 \cdots N_n$.

Digression on the Theory of Convex Functions

The general theory of convex functions and convex bodies is important for understanding how the qualitative features of phase diagrams are related to thermodynamics and statistical mechanics. In the spirit of the above review of the origin of present day definitions in thermodynamics, it is of interest to see to what extent the mathematical developments were influenced by problems of physics and vice-versa.

Of course, convex bodies and convex functions played an important role in mathematics and, in particular, in geometry long before they were studied in generality. The first general definition of convex functions on R^n appears to be that of Jensen [29]. He used a restricted definition of convexity

$$f\left(\frac{x+y}{2}\right) \leq \frac{1}{2}\,[f(x) + f(y)]\,. \tag{44}$$

He showed that in one dimension,

$$f(\alpha x + (1-\alpha)y) \leq \alpha\, f(x) + (1-\alpha)\, f(y) \tag{45}$$

for α rational and $0 \leq \alpha \leq 1$, follows from this, and went on to prove (45) for α real, under the assumption that f is bounded. Under this same assumption, he also proved f continuous. It had already been proved earlier that if f is convex in the customary sense that (45) holds for α real, $0 \leq \alpha \leq 1$, then f has left and right derivatives at each point of any interval on which it is defined [30]. Jensen also proved what is now called Jensen's inequality in several forms. For example, he showed

$$\phi\left(\frac{\sum_{\nu=1}^{n} a_\nu x_\nu}{\sum_{\nu=1}^{n} a_\nu}\right) \leq \frac{\sum_{\nu=1}^{n} a_\nu \phi(x_\nu)}{\sum_{\nu=1}^{n} a_\nu} \tag{46}$$

for ϕ convex in his sense with $a < x_1 < x_2, \cdots x_n < b$ and $a_1 \cdots a_n \geq 0$, $\sum_{\nu=1}^{n} a_\nu > 0$. He further showed that if f is convex and non-decreasing and ϕ is convex

$$f\left(\phi\left(\frac{x+y}{2}\right)\right) \leq \frac{1}{2}\left[f(\phi(x)) + f(\phi(y))\right]. \tag{47}$$

In a postscript, Jensen noted that Hölder had already proved the inequality (46) for a convex function, ϕ, under the assumption that ϕ exists and is bounded on subintervals of $a < x < b$. (Hölder's proof has no relation to the elegant derivation of Jensen's inequality from Hölder's inequality given in Israel's Chapter I!) [31].

Some of the most interesting applications of the theory of convex functions of that era were made by H. Minkowski, primarily to number theory and geometry. For an evaluation of the significance of this work, one can read Hilbert's memorial address [32]. For our present purposes, it is worth commenting on two of Minkowski's definitions. For each convex set, K, in a linear space, E, with 0 an interior point of K, Minkowski defined a *gauge function*, ρ_K, by

$$\rho_K(x) = \inf_{\substack{\lambda > 0 \\ x \in \lambda K}} \lambda . \tag{48}$$

(K is convex if $x, y \in K$ implies $\alpha x + (1-\alpha) y \in K$ for $0 \leq \alpha \leq 1$.) He showed that the properties of K can be expressed simply in terms of the properties of ρ_K [33]. For example, if $\rho_K(x) < 1$, x is an interior point of K; if $\rho_K(x) = 1$, x is a boundary point of K; if $\rho_K(x) > 1$, x lies in the exterior of K. Another example is the proposition: if F is a linear functional not identically zero on the linear space E, and H is the hyperplane defined by

$$F(x) = 1$$

then, for a convex set K to lie on one side of K and to have 0 as an interior point it is necessary and sufficient that

$$F(x) \leq \rho_K(x)$$

for all $x \in E$. The gauge function is not only a convex function, it is homogeneous of degree 1

$$\rho_K(\lambda x) = \lambda \rho_K(x) \qquad \text{for } \lambda > 0$$

and so is subadditive in the sense that

$$f(x+y) \leq f(x) + f(y) .$$

Thus, it is what nowadays is called a *semi-norm.*

Minkowski discussed the notion of gauge function in the context of three dimensional Euclidean space, but the definition makes sense for an arbitrary vector space, and later developments have shown its usefulness in that generality. In the now standard theory of locally convex topological vector spaces, there is a base for the neighborhoods of each point consisting of convex (symmetric, absorbing) sets, K. By introducing the corresponding gauge functions, ρ_K, one sees that the topology can be defined by a family of semi-norms [34].

Minkowski defined a second function associated with a convex set K, the *support function* $\rho^*(\ |K)$. It is a real (or $+\infty$) valued function defined on the dual, E^*, of the vector space E, by

$$\delta^*(x^*|K) = \sup_{x \in K} \langle x^*, x \rangle .$$

Here x^* is any continuous linear functional on E, i.e., any element of

E^*, and $\langle x^*, x \rangle$ is its value at the point x of E. In Euclidean space, the expression $\langle x^*, x \rangle$ can be taken as the Euclidean scalar product between the vector x^* specifying the linear functional and the vector x. Then one has the picture shown in Figure 19.

Minkowski proved that the support function of a convex set characterizes the set. This fact can be expressed quantitatively if one introduces the notion of indicator function $\delta(\cdot|K)$ of a convex set

$$\delta(x|K) = \begin{cases} 0 & x \in K \\ +\infty & x \notin K . \end{cases}$$

It then turns out that $\delta(\cdot|K)$ and $\delta^*(\cdot|K)$ are conjugate convex functions.

These remarks will perhaps give the flavor of Minkowski's work. If we ask what was its impact on physics, especially thermodynamics and statistical mechanics, the answer appears to be: nil. Indeed in the principal mathematical investigation of the foundations of thermodynamics of that time—that of Caratheodory, the connection of convexity with thermodynamic

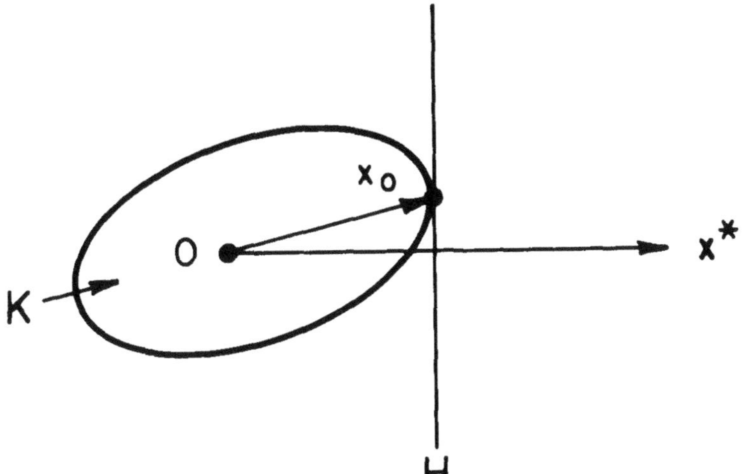

Fig. 19. *Geometrical meaning of the support function,* $\delta^*(\cdot|K)$, of a convex set K. $\langle x^*, x \rangle$ is the projection of x on the x^* direction multiplied by the magnitude of x^*. The support function gives the value of this quantity when $x = x_0$, where x_0 is a point at which the hyperplane, H, orthogonal to x^* and touching the set K, touches K. H is a tangent plane of K in the sense of Israel's book.

stability is not even mentioned [35]. Conversely, the concrete convex thermodynamic functions do not appear to have been the inspiration for any mathematical enterprises. It was a long time before mathematical physicists began again to take such matters seriously. (More will be said about that in the next section on statistical mechanics.) Long before the mathematical physicists, the mathematical economists absorbed the message of Minkowski, applied his ideas and went on to further mathematical developments [36].

I have already remarked in connection with the definition of the enthalpy and free energies (23), (24), (25) that Gibbs argued that the Legendre transforms of the particular convex functions he was dealing with, are concave, and that the general mathematical theorem involved was recognized very late, even for convex functions on $\mathbf{R}^n$ –it was, in fact, in 1948 that Fenchel proved [18]:

THEOREM. *Let* G *be a non-empty convex subset of* $\mathbf{R}^n$ *and* f *be a convex lower semi-continuous function defined on* G, *such that*

$$\lim_{x \to y} f(x) = +\infty$$

for each boundary point of G *which does not belong to* G. *To each such pair* $\{G, f\}$ *there exists a unique conjugate* $\{\Gamma, f^*\}$ *consisting of a subset* Γ *of* $\mathbf{R}^n$ *and a function* f^* *defined on* Γ *satisfying the same conditions as* $\{G, f\}$, *and in addition*

$$\langle x^*, x \rangle \leq f(x) + f^*(x^*) \qquad \text{for all } \begin{array}{l} x \in G \\ x^* \in \Gamma . \end{array}$$

If x *is an interior point of* G, *there exists at least one* x^* *such that equality holds.* $\{G, f\}$ *is also the conjugate of* $\{\Gamma, f^*\}$.

The requirement of lower semi continuity deserves some comment. (A function, f, is lower semicontinuous at a point, y, in its domain $D(f)$, if $\liminf_{x \to y} f(x) = \lim_{R \to 0} \inf_{\substack{x \in D(f) \\ |x-y| \leq R}} f(x) = f(y)$.) This requirement is included to make the conjugate of the conjugate of $\{G, f\}$ equal to $\{G, f\}$. At any interior point of G, a convex function, f, is continuous, hence surely lower semi-continuous. At a boundary point of G, where it happens to be defined f satisfies

$$\liminf_{x \to y} f(x) \leq \liminf_{\alpha \to 0} f(\alpha x + (1-\alpha)y) \leq f(y) .$$

Thus the functions admitted in the theorem are those for which equality holds in this inequality. If one is given a convex function for which strict inequality holds at some boundary points of its domain, one can modify its definition at boundary points so as to achieve lower semi continuity without destroying the convexity property.

A second comment: the theorem assumes that f takes finite values on G. An alternative treatment would admit the possibility that f takes the value $+\infty$. Then G is defined as the set on which f takes finite values. This form of the theorem has been used in proving that $\delta(\cdot|K)$ and $\delta^*(\cdot|K)$ are conjugate functions.

The usefulness of Fenchel's theorem in relating thermodynamic functions has already been made clear in connection with equations (28)...(31).

Another useful family of results concerns the relation of the points of a convex body to its extreme points.

THEOREM [37]. *A closed bounded convex set in* R^n *is the convex hull of its extreme points.*

By definition the convex hull of K is the set of all convex combinations of pairs of points in K: $\{\alpha x + (1-\alpha)y\,;\ 0 \leq \alpha \leq 1,\ x, y \in K\}$.

This theorem was already established by Minkowski. It is what one uses to express mixed thermodynamic phases in terms of pure phases.

The preceding discussion of convex functions and convex sets dealt almost entirely with functions defined on and sets lying in finite dimensional vector spaces. The generalization of the results to infinite dimensions turned out to be non-trivial. In fact, the results for the infinite dimensional case constitute some substantial chapters of modern functional analysis. Remarkably enough, infinite dimensional Banach spaces and their dual spaces of linear functionals turned out to be tailor-made for the study of general lattice systems. For the purposes of these applications the main general theorems can be classified as follows.

1) *separation theorems*

These are based on the Hahn-Banach theorem in one of its forms [38]. Among its corollaries is the statement that at each point $\{x, f(x)\}$ of the graph of continuous convex function, f, one can draw at least one *tangent hyperplane*. (That is, at least one surface of the form

$$y(x) = f(x_0) + \langle x^*, x-x_0 \rangle$$

such that the points of the graph of f lie above the hyperplane:

$$f(x) \geq y(x) = f(x_0) + \langle x^*, x-x_0 \rangle \quad \text{for all } x.)$$

(We follow the notation of Israel's book in using the word tangent. Elsewhere such a hyperplane is sometimes called a *supporting hyperplane* and the adjective tangent is reserved for the case in which it is unique.) These theorems were first proved in the 1920's for Banach spaces and in the 1930's for locally convex topological spaces.

2) *generation theorems*

These theorems tell how the points of a compact convex set, K, are obtained as barycenters of the extreme points of K. The refined form of the theory, customarily referred to as Choquet theory, gives an integral representation in terms of a measure concentrated on the extreme points. It also gives conditions for the integral representation to be unique [39]. Choquet theory dates from the late fifties and early sixties.

3) *category theorems on the existence of tangent planes*

Here Mazur's theorem says that the points of the graph of a continuous convex function at which the tangent plane is not unique are rare [40]. In the application it may be interpreted as a very gross form of the Gibbs Phase Rule.

Israel's Chapter VI contains a remarkable refinement of the theorem of Mazur. He introduces a definition of a ℓ-dimensional linear subspace of interactions satisfying the Gibbs Phase Rule: the subsets of those interactions for which k or more coexisting phases is empty for $k > \ell$ and has Hausdorff dimension $\ell - k$ for $k \leq \ell$. He then proves that in an appropriate sense the family of those linear subspaces that satisfy the Gibbs Phase Rule is a countable intersection of dense open sets.

4) *approximation theorems for tangent planes*

Mazur's theorem gives one the impression that one should be able to get the tangent planes at exceptional points as limits of those at points where the tangent plane is unique. The theorem of Lanford and Robinson says that all tangent planes are in the closed convex hull of those tangent planes that are limits of sequences of tangent planes that are unique at their points of tangency [41]. The theorem of Israel describes how P-bounded linear functionals can be approximated by functionals tangent to P. More explicitly, given a convex lower semi-continuous function on a Banach space, B, one says that a continuous linear functional a on B is P-*bounded* if there exists a constant C such that

$$P(\Phi) \geq a(\Phi) + C \quad \text{for all } \Phi \in B .$$

The theorem asserts that, given a $\Phi_0 \in B$ such that $\Phi(\Phi_0) < \infty$ and a P-bounded continuous linear functional a_0, there exists a point $\tilde{\Phi} \in B$ and a continuous linear functional $\tilde{a}$ tangent to P at $\tilde{\Phi}$ such that $\tilde{a}$ is close to a_0 in norm and $\|\Phi - \Phi_0\|$ has an explicitly given bound. Israel's theorem is a refinement of a theorem of Bishop and Phelps [42]. It is described in Chapter V of Israel's book, where some remarkable physical conclusions are drawn from it.

To make these statements somewhat more concrete I give the statement of two specific theorems, one of which is a generation theorem, the other a category theorem.

Although in infinite dimensional topological vector spaces closed bounded convex sets need not have any extreme points at all, [43] compact sets have extreme points. In fact, they have enough to make the following generalization of the preceding theorem true.

THEOREM (Krein-Milman) [44]. *In a locally convex topological vector space every compact convex set is the closed convex hull of its extreme points.*

(Recall that a compact set, K, is one such that if $\{U_\alpha, \alpha \in I\}$ is a family of open sets which covers K, then there is a finite sub family $\{U_\alpha ; \alpha \in I'\}$ which also covers K. In $\mathbf{R}^n$ every closed bounded set is compact, but that is not true in an infinite dimensional Banach space.) Notice that it is necessary, in general, to take the closure of the convex hull of the extreme points, unlike the finite dimensional case.

For a convex function of a single variable left and right derivatives exist at each point and are equal except at a denumerable set of points. Thus the set of points where a convex function of single variable does not have a unique tangent is very sparse. A generalization of this result to Banach spaces is

THEOREM (Mazur) [40[. *A convex function on a separable Banach space has a unique tangent plane on a set which contains a countable intersection of dense open sets, i.e., a dense* G_δ.

Sets G_δ are "large" in a rather weak sense. That the Banach space is separable means that it has a dense sequence of points.

Early Statistical Mechanics

One of the great problems of equilibrium statistical mechanics is to show how the laws of thermodynamics arise from microscopic dynamics. The issue here is not an analysis of the approach to equilibrium of non-equilibrium states–that is another great problem and one whose solution is necessary for a full understanding of the notion of equilibrium–but of finding a definition of equilibrium state and of the associated thermodynamic functions in terms of the Hamiltonian so that the laws of thermodynamics follow. Pioneering work on this problem was done by Maxwell and Boltzmann. Their ideas were codified by Gibbs whose formulation of equilibrium statistical mechanics has persisted to this day in classical statistical mechanics and was adopted with only a few formal changes in quantum statistical mechanics, once quantum mechanics was invented [45].

Given a Hamiltonian system of N degrees of freedom with Hamiltonian, H_N, Gibbs defined two probability measures on phase space, the microcanonical and canonical ensembles. If, for simplicity the phase space is assumed to be $\mathbf{R}^{2N}$, the *microcanonical ensemble* for a system of energy U in a subset V of $\mathbf{R}^n$ is defined by the probability measure

$$d\,\mu_{\mathrm{micro}}(U,V,N;\,q,p) = [Z_{\mathrm{micro}}(U,V,N)]^{-1}\,\frac{1}{N!}\,\chi_V(q)\,\delta(U-H_N)\,d^N q\;d^N p \tag{49}$$

where the normalization factor $Z_{\mathrm{micro}}(U,V,N)$ is the *microcanonical partition function*

$$Z_{\mathrm{micro}}(U,V,N) = \frac{1}{N!}\int \delta(U-H_N)\,\chi_V(q)\,d^N q\;d^N p\,. \tag{50}$$

Here

$$d^N q = dq_1 \cdots dq_N \qquad\qquad d^N p = dp_1 \cdots dp_N$$

δ is Dirac's delta function and $\chi_V(q) = 0$ if $q \notin V$ while it is $=1$ if $q \in V$. Gibbs also considered the possibility of using some other function in place of Dirac's δ, say the characteristic function of an interval; in the limit of a large system these possibilities turned out to be equivalent.

The *canonical ensemble* for a system at temperature T in a subset V of $\mathbf{R}^n$ is defined by the probability measure

$$d\mu_{\text{can}}(\beta,V,N;q,p) = Z_{\text{can}}(\beta,V,N)^{-1}\frac{1}{N!}\exp[-\beta H_N]\,\chi_V(q)\,d^Nq\,d^Np \tag{51}$$

where $\beta = (kT)^{-1}$ and the normalization factor $Z_{\text{can}}(\beta,V,N)$ is the *canonical partition function*

$$Z_{\text{can}}(\beta,V,N) = \frac{1}{N!}\int \exp[-\beta H_N]\,\chi_V(q)\,d^Nq\,d^Np\,. \tag{52}$$

The third ensemble used by Gibbs is especially adapted to systems containing arbitrary numbers of identical particles. Then it is natural to adapt the coordinates and momenta appearing in the Hamiltonian to the particles, using N as a label for the number of particles. The phase space is then $\mathbf{R}^{6N}$, and V refers to a region in $\mathbf{R}^3$. Given to a sequence of Hamiltonians H_N, $N = 1,2,3\cdots$ one then defines a measure on the product space $\mathop{\mathrm{X}}\limits_{N=0}^{\infty} \mathbf{R}^{6N}$, by

$$d\mu_{\text{grand}}(\beta,V,\mu;qp) =$$

$$[Z_{\text{grand}}(\beta,V,\mu)]^{-1}\sum_{n=0}^{\infty}\frac{1}{n!}\exp[-\beta H_n+\beta\mu n]\,d^{3n}q\,d^{3n}p\,. \tag{53}$$

Here on the left hand side qp is a point of $\mathop{\mathrm{X}}\limits_{N=0}^{\infty} \mathbf{R}^{6N}$, i.e., a sequence $0,0;\ \vec{q}_1^{(1)}\,\vec{p}_1^{(1)};\ \vec{q}_1^{(2)}\,\vec{p}_1^{(2)}\,\vec{q}_2^{(2)}\,\vec{p}_2^{(2)};\cdots$ and by convention $\mathbf{R}^0$ is a one point space with measure 1. This probability measure is called the *grand canonical ensemble*. The parameter μ is the chemical potential. The normalization factor $Z_{\text{grand}}(\beta,V,\mu)$ is the *grand canonical partition function*

$$Z_{\text{grand}}(\beta,V,\mu) = \sum_{n=0}^{\infty} e^{\beta\mu n}\,Z_{\text{can}}(\beta,n,V)\,. \tag{54}$$

For each of these ensembles Gibbs defined a corresponding thermodynamic function

$$S_{\text{micro}}(U,V,N) = \ell n\, Z_{\text{micro}}(U,V,N) \tag{55}$$

$$-\beta A_{\text{can}}(\beta,V,N) = \ell n\, Z_{\text{can}}(\beta,V,N) \tag{56}$$

$$\cdot\beta|V|p_{\text{grand}}(\beta,V,\mu) = \ell n\, Z_{\text{grand}}(\beta,\mu,V)\,. \tag{57}$$

Each of these functions S_{micro}, A_{can}, p_{grand} is a candidate for a thermodynamic function defining a fundamental equation in the sense of the definition after equation (8). Thus, starting from the microcanonical entropy, S_{micro}, one can compute micro-canonical Helmholtz free energy A_{micro}. In general A_{micro} and A_{can} will not be equal. Gibbs recognized that it was only by passing to a limit in which the number of degrees of freedom $N \to \infty$ and the volume $|V| \to 0$ with the density $N/|V|$ approaching a finite limit that one could expect the fundamental equations calculated from the different ensembles to agree. He says [46]

> "A very little study of the statistical properties of conservative systems of a finite number of degrees of freedom is sufficient to make it appear, more or less distinctly, that the general laws of thermodynamics are the limit toward which the exact laws of such systems approximate when their number of degrees of freedom is indefinitely increased."

A more explicit version of these statements is this: the three ensembles ought to define, respectively, an entropy per unit volume

$$s(u,v) = \lim_{V\to\infty} \frac{1}{|V|} S_{\text{micro}}(U,N,V) \tag{58}$$

as a function of the energy density $u = \lim_{V\to\infty} \frac{U}{V}$ and specific volume $v = \lim_{V\to\infty} \frac{V}{N}$, a free energy per unit volume

$$a(\beta,v) = \lim_{V\to\infty} \frac{1}{|V|} A_{\text{can}}(\beta,N,V) \tag{59}$$

as a function of β and the specific volume and a pressure

$$p(\beta,\mu) = \lim_{V\to\infty} p_{grand}(\beta,\mu,V) \tag{60}$$

as a function of β and the chemical potential μ. These three limit functions ought to be consistent with one another in that they define the same thermodynamic fundamental equation. The limit indicated formally here by $V \to \infty$ implicitly assumes that $\frac{|V|}{N} \to v$ and that there is some restriction on the shape of V as $|V| \to \infty$ so that the limit exists. It is nowadays customary to refer this as the *thermodynamic limit*. This terminology is perhaps unfortunate since, as Gibbs also discussed, the thermodynamics of finite systems also makes sense provided, for the canonical and grand canonical ensembles, that they are in contact with an appropriately defined heat reservoir. The differences between ensembles are then accounted for by the different nature of the interaction of the finite system with the reservoir. (For the canonical ensemble only heat is exchanged with the reservoir, whereas for the grand canonical ensemble heat and particles are exchanged.) [47] We can summarize the situation by saying that the three ensembles are expected to be equivalent in the thermodynamic limit.

Gibbs did not attempt a direct mathematical proof of the existence of the thermodynamic limit although he did give plausible arguments that fluctuations in thermodynamic quantities would tend to decrease in the thermodynamic limit. In fact, more than forty years passed before van Hove and Lee and Yang gave the first mathematical proofs [48][49] and more than fifty before a full scale attack on the problem for various kinds of systems was undertaken at the beginning of the development of modern statistical mechanics [50]. The idea that one should study systematically for what class of Hamiltonians the limit exists and for what it fails to exist, simply was not given a high priority among mathematical physicists during the first half of the twentieth century. Of course, in a mathematically rigorous treatment of statistical mechanics such a study is indispensible and, as one can see from hindsight, it turned up some useful physical discoveries.

An example is the notion of *stability* for a two-body potential $V(r)$. Contrary to what one might think, it is not enough that a potential should have a repulsive core to guarantee that the thermodynamic limit should exist. See Figure 20.

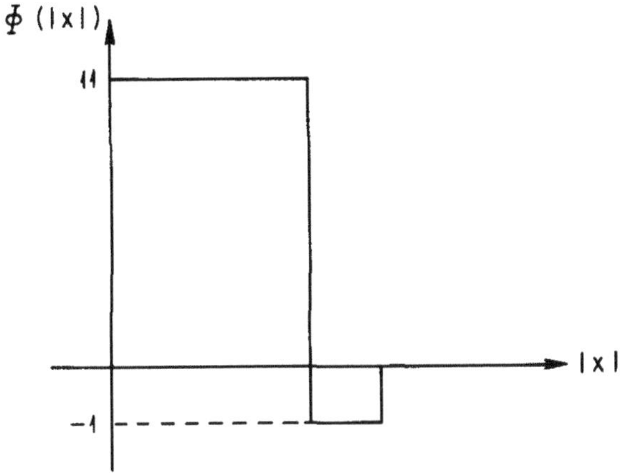

Fig. 20. A two-body potential for which macroscopic matter would not exist [51].

Gibbs offered several arguments in favor of the thermodynamic analogy, as he called it, i.e., the compatibility of the expressions (58) (59) (60) with the laws of thermodynamics, but nothing that would be recognized nowadays as a proof. As an example of the state of the subject in the first decade of the twentieth century consider the following theorem and the use Gibbs made of it. For an ensemble of Hamiltonian systems of N degrees of freedom let

$$P(p_1 q_1, \cdots, p_n q_n)\, d^N p\, d^N q \tag{61}$$

be the probability distribution in phase space. Here P is a positive function such that

$$\int P(p_1 q_1 \cdots p_n q_n)\, d^N p\, d^N q = 1 . \tag{62}$$

Gibbs writes $P = \exp \eta$ and calls η the index of probability. He proves [52]

"*Theorem VII.* If a system which in its different phases constitutes an ensemble consists of two parts, and we consider the average index of probability for the whole system, and also the average indices for each of the parts taken separately, the sum of the average indices for the parts will be either less than the average index for the whole system, or equal to it, but cannot be greater. The limiting case of equality occurs when the distribution in phase of each part is independent of that of the other, and only in this case."

More explicitly, suppose the subsystems have the coordinates $p_1 q_1 \cdots p_n q_n$ and $p_{n+1} q_{n+1} \cdots p_N q_N$ respectively. If η_1 and η_2 are defined by

$$\int \cdots \int e^{\eta} \, dp_{n+1} \, dq_{n+1} \cdots dp_N \, dq_N = e^{\eta_1}$$

$$\int \cdots \int e^{\eta} \, dp_1 \, dq_1 \cdots dp_n \, dq_n = e^{\eta_2}$$

then the assertion of the theorem is

$$\overline{\eta}_1 + \overline{\eta}_2 \leq \overline{\eta} \tag{63}$$

where

$$\overline{\eta}_1 = \int \eta_1(p_1 q_1 \cdots p_n q_n) \, e^{\eta_1} \, dp_1 \, dq_1 \cdots dp_n \, q_n$$

$$\overline{\eta}_2 = \int \eta_2(p_{n+1} q_{n+1} \cdots p_N q_N) \, e^{\eta_2} \, dp_{n+1} \, dq_{n+1} \cdots dp_N \, dq_N$$

$$\eta = \int \eta(p_1 q_1 \cdots p_N q_N) \, dp_1 \, dq_1 \cdots dp_N \, dq_N \, .$$

Gibbs' proof of the theorem is impeccable. At first sight it seems to be just what is needed to prove that the entropy obtained from the canonical ensemble satisfies the maximum principle (16), from which the concavity of the entropy was derived in (17). One has only to write

$$S_{can}^{(1)} = -\eta_1 \qquad S_{can}^{(2)} = -\eta_2$$

$$S_{can} = -\eta \, .$$

Then the inequality (63) is

$$\overline{S}_{can} \geq \overline{S}_{can}^{(1)} + \overline{S}_{can}^{(2)} \, . \tag{64}$$

Why does this inequality not suffice to prove the required concavity? The answer is that the thermodynamic derivation takes for granted that the entropy is an extensive quantity, a property which is false for the expressions appearing in (64), though it is guaranteed in the thermodynamic limit, if the thermodynamic limit exists. (The entropy is then by definition $S(U,N,V) = VS\left(\frac{U}{V}, \frac{V}{N}\right)$ where S is the entropy per unit volume given by (58).) Thus, the problem of deducing the concavity of the canonical entropy from (64) is intimately connected with the existence of the thermodynamic limit itself [72].

There is one family of convexity properties which holds even in a finite volume. That is the convexity of the logarithm of the partition function $\ell n\ Z$ as a function of any parameters that appear linearly in the exponential under the integral sign in Z; this is a simple consequence of Hölder's inequality. For example, if

$$Z(\alpha) = \int e^{\alpha \cdot f(x)} \, d\mu(x) \tag{65}$$

where α stands for $\alpha_1 \cdots \alpha_k$ and $\alpha \cdot f(x)$ for $\sum_{i=1}^{k} \alpha_i f_i(x)$ then, if λ satisfies $0 < \lambda < 1$

$$\begin{aligned} Z(\lambda\alpha + (1-\lambda)\beta) &\leq \left[\int [e^{\lambda\alpha \cdot f(x)}]^{1/\lambda} \, d\mu(x)\right]^{\lambda} \times \\ &\qquad \left[\int [e^{(1-\lambda)\beta \cdot f(x)}]^{\frac{1}{1-\lambda}} \, d\mu(x)\right]^{(1-\lambda)} \\ &= \quad [Z(\alpha)]^{\lambda} \, [Z(\beta)]^{(1-\lambda)} \end{aligned}$$

i.e.,

$$\ell n\ Z(\lambda\alpha + (1-\lambda)\beta) \leq \lambda\ \ell n\ Z(\alpha) + (1-\lambda)\ \ell n\ Z(\beta)\ . \tag{66}$$

This accounts for the fact that $\ell n\ Z_{\text{can}}(\beta,N,V)$ is convex in β and $\ell n\ Z_{\text{grand}}(\beta,\mu,V)$ in β and $\beta\mu$ even before the passage to the thermodynamic limit. This remark is used systematically in the theory of lattice systems.

It is curious that, in his treatise on statistical mechanics, Gibbs chose *not* to emphasize the following points, especially since his first two papers on thermodynamics gave such a simple and precise description of thermodynamic stability and phase transitions in terms of the geometrical properties of convex functions.

a) Because the partition functions for systems with a finite number of degrees of freedom depend analytically on the parameter β it is to be expected that the sharp discontinuities or infinities at phase transitions in such quantities as specific heats will only appear in the thermodynamic limit.

b) Once the required convexity of the thermodynamic functions in the thermodynamic limit has been proved starting from a given Hamiltonian, the theory of phase transitions, Maxwell's Rule... would follow. One has only to look for the flat places in the graph of an appropriate thermodynamic function to find the values of the thermodynamic function which correspond to coexisting phases. In principle, the formalism should yield solid, liquid, and gas phases.

It is interesting to speculate that if Gibbs had chosen to lay out the problems in such a fashion much general confusion would have been avoided in the later development of the theory of phase transitions. Perhaps, he was influenced by his concern about the fact that the classical theory of solids and specific heats was in poor agreement with experiment. He said on this subject [53] "... the electrical phenomena associated with the combination of atoms, seem to show that the hypothesis of systems of a finite number of degrees of freedom is inadequate for the explanation of the properties of bodies.

Nor do the results of such assumptions in every detail appear to agree with experience. We should expect, for example, that a diatomic gas, so far as it could be treated independently of the phenomena of radiation, or of any sort of electrical manifestations, would have six degrees of freedom for each molecule. But the behavior of such a gas seems to indicate not more than five.

But although these difficulties, long recognized by physicists,* seem to prevent, in the present state of science, any satisfactory explanation of the phenomena of thermodynamics as presented to us in nature, the ideal case of systems of a finite number of degrees of freedom remains as a subject which is certainly not devoid of a theoretical interest, and which may serve to point the way to the solution of the far more difficult problems presented to us by nature. And if the study of the statistical properties of such systems gives us an exact expression of the laws which in the limiting case take the form of the received laws of thermodynamics, its interest is so much the greater. *See Boltzmann Sitz. der Wiener Akad LXIII (1871) 418."

After the appearance of Gibbs book on statistical mechanics a conceptual framework of classical equilibrium statistical mechanics was fixed that has not been altered in any essential way since: there is a class of dynamical systems which are thermodynamically stable in the sense that they possess a thermodynamic limit in which the laws of equilibrium thermodynamics hold. However, further studies in statistical mechanics led to a refinement of the notion of equilibrium state, and it is to this development which we now turn.

Refinement of the Notion of State in Statistical Mechanics

The striking alteration of the properties of a liquid in the neighborhood of the liquid-gas critical point is familiar to anyone who has taken a good elementary physics course. As the temperature of a critical volume of the liquid is raised, it bubbles wildly, becomes milky, and suddenly evaporates into a gas. The milkiness arises from density fluctuations. The theory of such fluctuations was one of the important achievements of statistical mechanics in the first twenty years of the twentieth century [54]. The theory describes the fluctuations in terms of a two point correlation function defined for the canonical ensemble by

$$<\rho(\vec{x})\,\rho(\vec{y})>_{\beta,V,N} = \int \left[\frac{1}{|V|}\sum_{j=1}^{N} \delta(\vec{x}-\vec{x}_j)\right]\left[\frac{1}{|V|}\sum_{k=1}^{N} \delta(\vec{y}-\vec{x}_k)\right] \tag{67}$$

$$d\mu_{can}(\beta, V, N;\, \vec{x}_1 \cdots \vec{x}_N, \vec{p}_1 \cdots \vec{p}_N$$

for a system of N particles of coordinates $\vec{x}_1 \cdots \vec{x}_N$ and momenta $\vec{p}_1 \cdots \vec{p}_N$. In the thermodynamic limit, this correlation function is predicted to have the cluster property for temperatures above the critical temperature, i.e.,

$$\lim_{x-y \to \infty} <\rho(x)\,\rho(y)>_\beta = <\rho(x)>_\beta <\rho(y)>_\beta\,. \tag{68}$$

Here

$$<\rho(x)>_\beta = \lim_{\substack{V \to \infty \\ N \to \infty}} \int \left[\frac{1}{|V|}\left(\sum_{j=1}^{N} \delta(\vec{x}-x_j)\right)\right] d\mu_{can}(\beta, V, N;\, \vec{x}_1 \cdots \vec{x}_N, \vec{p}_1 \cdots \vec{p}_N) \tag{69}$$

is the density of the gas (number of particles per unit volume); it is expected to be independent of x if the Hamiltonian of the system is translation invariant. For interactions of finite range, the approach to the limit is exponential in the sense that [73]

$$|<\rho(\vec{x})\,\rho(\vec{y})>_\beta - <\rho(\vec{x})>_\beta <\rho(\vec{y})>_\beta| \le C\, e^{-\frac{|\vec{x}-\vec{y}|}{\xi}}\,. \tag{70}$$

The correlation length, ξ, is a measure of how rapidly the correlation function approaches its limit as $|\vec{x}-\vec{y}| \to \infty$. ξ is a function of the temperature and, as the temperature approaches its critical value from above, it increases to $+\infty$. Later on it was recognized that scattering of neutrons or light by the system can also be expressed in terms of the two-point correlation function but, in general, with the times of the two points different

$$<\rho(\vec{x},t)\,\rho(y,t')>_{\beta,V,N} = \int \left[\frac{1}{|V|}\sum_{j=1}^{N}\delta(\vec{x}-\vec{x}_j(t))\right]\frac{1}{|V|}\left[\sum_{k=1}^{N}\delta(\vec{y}-\vec{x}_k(t')\right]$$

$$d\mu_{can}(\beta,V,N;\vec{x}_1\cdots\vec{x}_N\,\vec{p}_1\cdots\vec{p}_N)\,. \tag{71}$$

Here $\vec{x}_j(t)$ is the coordinate of the jth particle at time t when it starts from $\vec{x}_j(0)=\vec{x}_j$ at time $t=0$ with momentum $\vec{p}_j(0)=\vec{p}_j$ and moves according the equations of notion determined by the Hamiltonian H_N.

The two-point correlation function provides information about a system in equilibrium which is not directly given by the thermodynamic functions of the system. Thus the need for a detailed description of critical phenomena and of the scattering of light and neutrons from a system leads naturally to a refinement of the notion of state. Once one has admitted a single correlation function as part of the description of a state, it is difficult to avoid admitting them all, i.e., the thermodynamic limits of all of the following

$$<\rho(\vec{x}_1)\cdots\rho(\vec{x}_n)>_{\beta,V,N} = \int \prod_{k=1}^{n}\frac{1}{V}\left[\sum_{\ell=1}^{N}\delta(\vec{x}_k-\vec{y}_\ell)\right]$$

$$d\mu_{can}(\beta,V,N;\vec{y}_1\cdots\vec{y}_N,\vec{p}_1\cdots\vec{p}_N) \tag{72}$$

as n varies $n=1,2,\cdots$. This point of view was systematically developed by such workers as Yvon, Mayer, Montroll, Kirkwood, Salzburg, Born, and Green in the 1930's and 1940's. They found integral equations connecting the correlation functions and derived approximate solutions of them. Once one has accepted the refinement that an equilibrium state is given by a set of equilibrium correlation functions as well as a set of thermodynamic functions, one has to reexamine the conceptual framework which accompanies the formalism in which a state is specified by thermodynamic functions alone. Here are three examples of the problems that arise.

a) How should the notions of pure and mixed phase and that of equilibrium phase be extended?

b) How should invariance properties under the Euclidean group or its subgroups enter the definitions of pure and equilibrium phases?

c) The definitions of pure and equilibrium phases hold for the thermodynamic limit of a theory. Can all physically interesting quantities be evaluated on the actually infinite system obtained in this limit?

We make a long series of remarks bearing on these questions.

a) It is natural to extend to correlation functions the idea of making mixtures by taking convex linear combinations of thermodynamic functions: if $<\rho(\vec{x}_1)\cdots\rho(\vec{x}_n)>_\beta^{(1)}$ and $<\rho(\vec{x}_1)\cdots\rho(\vec{x}_n)>_\beta^{(2)}$ are, for $n = 1,2,\cdots$, the correlation functions of two equilibrium states, then the correlation functions of the mixtures one can obtain from these states are

$$\rho_{\beta,a}(\vec{x}_1\cdots\vec{x}_n) = a<\rho(\vec{x}_1)\cdots\rho(\vec{x}_n)>_\beta^{(1)} + (1-a)<\rho(\vec{x}_1)\cdots\rho(\vec{x}_n)>_\beta^{(2)} \quad (73)$$

for $0 < a < 1$. Of course, the $\rho_{\beta,a}$ may or may not be the correlation functions of some equilibrium state. When they are, for some a, then that equilibrium state is said to be a mixture of the states labeled by $^{(1)}$ and $^{(2)}$. If the correlation functions of an equilibrium state cannot be decomposed in this way for some a satisfying $0 < a < 1$, then the equilibrium state is said to be *extremal*.

These definitions seem reasonable, although they leave open the question how equilibrium state is to be defined. However, they are, in fact, quite subtle as one sees if one attempts to describe the behavior of a sequence of states of larger and larger finite systems such that, in the thermodynamic limit, they converge to such a mixed state of the actually infinite system. If, for concreteness, one thinks of a mixture of liquid and gas pure phases, then the large system has to consist mainly of a large droplet of liquid and the rest gas in proportions a and $(1-a)$. However, one also expects a finite probability for a more complicated distribution of droplets and it is the asymptotic probability distribution of the droplets as the system becomes larger and larger that determines the thermodynamic limit is a mixture.

How can one tell from a set of equilibrium correlation functions that a state is extremal? Interestingly enough, it was apparently Norbert Wiener who suggested an answer: extremal states have the cluster property: [55]

$$\lim_{a\to\infty} [<\rho(\vec{x}_1)\cdots\rho(\vec{x}_n)\,\rho(\vec{y}_1+\vec{a})\cdots\rho(\vec{y}_m+\vec{a})>_\beta - <\rho(\vec{x}_1)\cdots\rho(\vec{x}_n)>_\beta <\rho(\vec{y}_1)\cdots\rho(\vec{y}_m)>_\beta] = 0 \quad (74)$$

which is a generalization of (68). It is easy to see that a mixture of two distinct states which themselves have the cluster property, cannot have the cluster property because this difference is then

$$a(1-a)\,[<\rho(\vec{x}_1)\cdots\rho(\vec{x}_n)>_\beta^{(1)} - <\rho(\vec{x}_1)\cdots\rho(\vec{x}_n)>_\beta^{(2)}]$$
$$[<\rho(\vec{y}_1)\cdots\rho(\vec{y}_m)>_\beta^{(1)} - <\rho(\vec{y}_1)\cdots\rho(\vec{y}_m)>_\beta^{(2)}]$$

which cannot be zero for all m, n and all $\vec{x}$'s and $\vec{y}$'s unless $a=0$ or 1. It is perhaps not so surprising that Norbert Wiener should have had this insight since he had been working with probability distributions for the histories of dynamical systems and with ergodic theory since the 1920's.

This simple criterion for the extremal property of states requires refinement when the phenomenon of symmetry breaking takes place. To discuss this we need to go into the description of the symmetry of states in some detail.

b) There is a second distinguishing property of most of the equilibrium states discussed above: their invariance under spatial symmetry transformations. In our discussion we simply took this invariance for granted. However, a correct definition of equilibrium state ought, in general, to yield both invariant equilibrium states and non-invariant. (And, in fact, the third paper of Gibbs already discussed numerous examples of non-invariant thermodynamic equilibrium states.) As an example, consider a simple model of a ferromagnet in which each site of the lattice $\mathbf{Z}^\nu = \{\{n_1 \cdots n_\nu\};\ n_j \text{ an integer for } j = 1 \cdots \nu\}$ is occupied by a spin taking two values, ± 1, say, and an interaction takes place between nearest neighbors tending to align their spins. Then for a finite system, say a cube, with $0 \le |n_j| \le N/2$ one can encourage the spins in one half the system to point up by fixing the spins on part of the boundary of the cube up. One can encourage those in the other half to point down by fixing the spins on the other half of the boundary down. See Figure 21 (a). (The Hamiltonian and further details of the model are given in (83) ff.) Now for all sufficiently high temperatures, in the thermodynamic limit boundary conditions don't matter; one will always get a unique equilibrium state invariant under translations of the lattice. For low temperatures there are two extremal equilibrium states invariant. For dimension $\nu = 2$, these are the *only* extremal equilibrium states. However, for $\nu \geq 3$, the above boundary conditions lead, in addition, to extremal equilibrium states not invariant under translations in the left right direction [56]. See Figure 21 (b) for a schematic indication of their character. They are distinguished by a plane

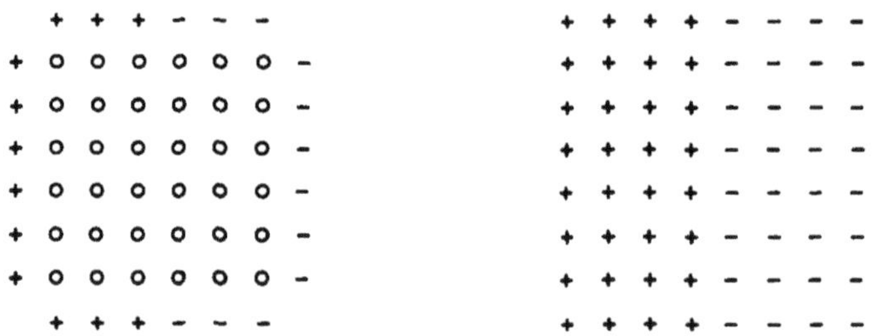

(a) (b)

Fig. 21. (a) Boundary conditions for producing a non-invariant equilibrium state for a Lenz-Ising ferromagnet in the plane. The circles represent lattice sites; the phases sites where the spin is up, the minuses sites where the spin is down.

(b) Schematic description of an extremal non-invariant equilibrium phase (only occurs for dimension, $\nu > 2$). The plus signs here indicate spins which are mainly up, the minus signs spins mainly down.

across which there is an inhomogeneity: or one side the spins are mainly up, on the other side, mainly down. There is evidently one such state for each dividing plane and another infinite such set with up spins interchanged with down spins. One current usage in statistical mechanics seems to be to regard these inhomogeneous equilibrium states as hybrids and to use the phrase *pure phase* only for translation invariant invariant extremal equilibrium states [57]. An alternative is to define the notion of pure phase relative to an invariance group: an equilibrium state is a pure phase relative to a group G, if it is extremal among G invariant equilibrium states. This latter definition makes the extremal equilibrium states of Figure 21 (b) pure phases also but relative to the translation group of the hyperplanes parallel to the inhomogeneity [58].

As an illustration of the consequences of accepting the refined notion of equilibrium state let us reconsider the example described in Figure 14 of a composite system with two non-interacting subsystems. We now choose these subsystems more specifically as two dimensional Lenz-Ising ferromagnets [20]. With this interpretation the V_1 and V_2 which we wrote before are replaced by, say, magnetizations per unit volume

$$m_1 = \langle \sigma_j^{(1)} \rangle_\beta \qquad m_2 = \langle \sigma_j^{(2)} \rangle_\beta$$

and the region of the phase transition is centered around the origin. Otherwise, the phase diagram in the thermodynamic variables remains qualitatively the same. However, the equilibrium states which were formerly labeled by T, V_1, V_2 and the thermodynamic function F now are labeled by spin correlation functions. Consider, in particular, the square of Figure 14; it violates the Gibbs Phase Rule by having its mid point one half the sum of two distinct pairs of pure thermodynamic phases, those represented by opposite corners

$$\frac{1}{2}[P+R] = \frac{1}{2}[Q+S] .$$

On the other hand, the spin correlation functions belonging to the four corners of the square are linearly independent and so form a tetrahedron in an appropriate space. Thus when equilibrium state is interpreted in the sense of statistical mechanics, the points represented by the square are to be interpreted as a quadruple point and no longer violate the Phase Rule. What does constitute such a violation is the fact that as the temperature, T, varies, there is a family of quadruple points forming a line. (The Gibbs Phase Rule would say that in a system with two components the set of quadruple points has dimension 0 not 1.)

c) The definition of equilibrium state in terms of correlation functions suffers from several defects having to do with a lack of completeness.

First, it may happen that information gets lost in the passage to the thermodynamic limit. For example, in the thermodynamic limit of the two-dimensional Lenz-Ising ferromagnet, there are no extremal equilibrium states not invariant under the translation group of the lattice. Thus, there are no equilibrium states in which it makes sense to discuss the "surface tension" of an inhomogeneity. On the other hand, there is an approximate surface tension associated with the inhomogeneity shown in Figure 21(a) and, as was already shown by Onsager in 1944, this approximate surface tension converges to a non-vanishing limit as the size of the system increases. Thus, there is information in the properties of the equilibrium states of the family of finite systems which is not directly expressible in terms of the equilibrium states of an actually infinite system. What is to be done? The answer accepted at the moment appears to be: work with the actually infinite system as much as possible, but when necessary adjoin information arising from finite systems.

Second, there may be incompleteness in the description in terms of correlation functions that arises from the unboundedness of the quantities ρ whose correlation functions are used [59]. This is a familiar phenomenon in problems that involve deducing a unique measure ρ from its moments. Consider, for example, a positive measure ρ on the real line with moments

$$\mu_n = \int \xi^n \, d\rho(\xi) . \tag{75}$$

The moments always satisfy

$$\sum \bar{c}_n c_m \mu_{n+m} \geq 0 \tag{76}$$

where the c_n are any finite set of complex numbers because the left hand side is

$$\int \left| \sum c_n \xi^n \right|^2 d\rho(\xi) .$$

Conversely, given a sequence of real numbers μ_n, $n = 0, 1, \cdots$ satisfying (76), there exists a measure ρ such that (75) holds. However, in general, ρ is not uniquely determined. To get uniqueness it suffices to impose some growth conditions on the μ_n as a function of n but a standard alternative procedure is to assume $\int f(\xi) \, d\rho(\xi)$ given for some more general class of f than polynomials. It is the latter which was developed in the context of statistical mechanics and quantum field theory under the general title: the C^* algebra approach. Most of the general theory of rigorous statistical mechanics is now written in terms of this language. What will be needed here may be summarized briefly as follows.

To each bounded region of space $\mathcal{O}$ (or, in a lattice system, to each finite subset of the lattice) there is associated an algebra generated by observables, $\mathfrak{A}(\mathcal{O})$. In the classical case, it is an abelian algebra of complex valued functions, the subset of real functions representing observables. In quantum mechanics, it is an algebra of bounded operators of which the hermitean operators are the observables. It is supposed that

$$\mathcal{O}_1 \subseteq \mathcal{O}_2 \text{ implies } \mathfrak{A}(\mathcal{O}_1) \subseteq \mathfrak{A}(\mathcal{O}_2)$$

and in the quantum mechanical case $\mathcal{O}_1$ and $\mathcal{O}_2$ non-overlapping implies $\mathfrak{A}(\mathcal{O}_1)$ and $\mathfrak{A}(\mathcal{O}_2)$ commute. The $\mathfrak{A}(\mathcal{O})$ are supposed to be normed algebras such that

$$\|A^*A\| = \|A\|^2 .$$

(This is a mathematical assumption which receives its justification when the theory turns out to be general and applicable to many important situations.) The closure in the norm of the algebra $\bigcup_{\mathcal{O}} \mathfrak{A}(\mathcal{O})$ is, $\mathfrak{A}$, the *quasi-local algebra*.

A *state* is defined as a positive linear form, σ, on the quasi-local algebra normalized to 1 at observable, 1:

$$\sigma(A^*A) \geq 0, \qquad \rho(1) = 1 .$$

A state can be thought of as an expectation value functional and as such is obviously related to the correlation functions previously employed. In fact, for spin correlation functions products such as

$$\sigma_{j_1} \cdots \sigma_{j_n}$$

are contained in the quasi-local algebra so a state in the C^* algebra sense does define a set of correlation functions. Unbounded functions like the $\rho(x_1) \cdots \rho(x_n)$ that appears in (72) are not in the quasi-local algebra but bounded functions of the ρ's are, so the C^* algebra notion of state includes that of correlation function in a heuristic sense at least.

The elegance and simplicity of the C^* algebra formalism shows itself in the general theory of states. For example, since a state is a positive linear form on a C^* algebra, it turns out to be automatically continuous in the sense that

$$|\sigma(A)| \leq \|A\|$$

where $\|\cdot\|$ is the norm in the quasi-local algebra $\mathfrak{A}$. The convex combinations $a\sigma^{(1)} + (1-a)\sigma^{(2)}$, $0 \leq a \leq 1$ are states, if $\sigma^{(1)}$ and $\sigma^{(2)}$ are states, so the notions of *extremal state and mixed state* can be defined in the usual way and have the physical significance we discussed in connection with correlation functions. According to a general theorem on Banach spaces the set of states on $\mathfrak{A}$ forms a compact set in A^*, the dual of $\mathfrak{A}$ regarded as a Banach space. Thus the Krein-Milman theorem which says that the set is the closure of the convex hull of its extreme points is applicable. The transformation properties of states under Euclidean transformations (or, for lattice systems, Euclidean transformations leaving the lattice invariant) is expressed in terms of the action of Euclidean transformations on the quasi-local algebra:

$$\sigma_{\{\vec{a},R\}}(A) = \sigma(\{\vec{a},R\}^{-1}A) \qquad \text{for } A \in \mathfrak{A} . \tag{77}$$

Here $\sigma_{\{\vec{a},R\}}$ is the state σ transformed by $\{\vec{a},R\}$, the Euclidean transformation which is the rotation R followed by the translation $\vec{a}$. The action of Euclidean transformations on elements of the quasi-local algebra is given by

$$(\{\vec{a},R\}A)(x) = A(R^{-1}(\vec{x}-\vec{a})) \tag{78}$$

if A is a classical scalar observable. In the quantum mechanical case we suppose there is a unitary operator $V(\vec{a},R)$ mapping the Hilbert space of $\mathfrak{A}(\mathcal{O})$ on to that of $\mathfrak{A}(R\mathcal{O}+\vec{a})$ such that

$$\{\vec{a},R\}A = V(\vec{a},R)A\,V(\vec{a},R)^* . \tag{79}$$

A state is *invariant* under $\{\vec{a},R\}$ if

$$\sigma_{\{\vec{a},R\}} = \sigma . \tag{80}$$

It is called G-*invariant* if it is invariant under a group, G, of such transformations.

There is a beautiful general theory of G-invariant states [60] which assures us, when G is the group of translations of $\mathbf{R}^\nu$, or those, $\mathbf{Z}^\nu$, of the lattice, that

1) G-invariant states form a simplex. Thus any G-invariant state can be written as a barycenter of extremal G-invariant states in an essentially unique way.

2) Extremal G-invariant states form pure phases in the sense that the fluctuations in averaged quantities vanish in the limit of large volume

$$\lim_{V\to\infty} \rho\left(\left[\frac{1}{|V|}\int d\vec{a}\{\vec{a},1\}A - \rho(A)\right]^2\right) = 0 . \tag{81}$$

3) Extremal G-invariant states are characterized by cluster properties.

This theory gives a general framework for a discussion of *symmetry breaking* [61]. It may happen that a state, ρ, is G-invariant and extremal among G-invariant states but that for a subgroup, H, of G it is not extremal among H-invariant states. The decomposition of ρ into H-invariant states is said to be an instance of symmetry breaking. An example occurs for equilibrium states of the Lenz-Ising model with anti-ferromagnetic interaction in which nearest neighbors tend to be oppositely aligned. At sufficiently low temperatures, the spins on one sublattice Λ_1 tend to be aligned one way and those on the complementary lattice Λ_1^c the opposite way. There are two extremal equilibrium states at such temperatures which are invariant under the translation group with double

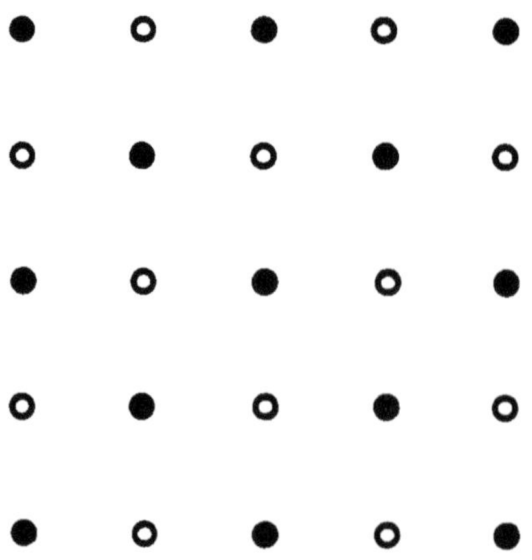

Fig. 22. Complementary sublattices for the Lenz-Ising anti-ferromagnet. Black spots o, which form the sublattice, Λ_1, and light spots o, which form its complement, Λ_1^c, tend to have oppositely aligned spins.

spacing, $2\,\mathbf{Z}^2$, but not under the translation group $\mathbf{Z}^2$ itself. The equilibrium phase which is a mixture of the two extremal phases in equal proportions *is* $\mathbf{Z}^2$ invariant. The general theory of such symmetry breaking for the Euclidean group was developed in [61]. It shows how a theory of the classification of crystals can be given within this formalism.

The general theory of G-invariant states is purely kinematical. It makes no mention of any particular microscopic dynamics. To arrive at a general definition of equilibrium state one has specify the dynamics by one means or another. There are three general approaches which have been studied. They may be labelled

1) Equilibrium Equations.
2) Variational Equations.
3) Tangents to the Pressure.

Since each of these starts from the same heuristic ideas, it is not unreasonable to expect that they define the same equilibrium states for well behaved systems. On the other hand, one should not be surprised to find that pushed to their limits on malignant interactions they can be defined for somewhat differing classes.

The equilibrium equations for a classical system express the fact that an equilibrium state of an infinite system should reduce to a grand canonical ensemble when attention is confined to a finite volume. For a quantum system the analogous requirements involve the KMS (Kubo-Martin-Schwinger) condition. On the other hand, the variational equation says that equilibrium states may be characterized as those which yield equality in the variational principle for the pressure

$$\beta p = \sup_{\rho} \left[s(\rho) - \beta u(\rho) \right] . \tag{82}$$

Here the sup is taken over all states ρ of finite entropy per unit volume $s(\rho)$. $u(\rho)$ is the energy per unit volume. (The variational principle is just the statistical mechanical version of the thermodynamic equation (31), with the contributions from the chemical potentials included, by definition, in the energy $u(\rho)$.) Finally, tangents to the pressure realize the idea that the correlation functions should be expressible as functional derivatives of thermodynamic functions with respect to interaction constants or potentials.

The systematic development of the equivalence of these approaches is far from complete for continuous systems, but there is every indication that the conceptual framework is sound. For classical and quantum lattice systems, on the other hand, the general theory goes very far as the book of Israel shows, and it has become possible to consider more detailed and physical questions.

Lattice Systems; A Testing Ground

The statistical mechanics of lattice systems has played an important role in clarifying some of the general issues of statistical mechanics. As a preliminary to the discussion of the general theory of such systems, here is a little more information on the Lenz-Ising model.

The model associates a two-valued spin, $\sigma_2 = \pm 1$, with each point i of the lattice $\mathbf{Z}^\nu$, $\mathbf{Z}^\nu$ being the set of points with integer coordinates in ν-dimensional Euclidean space. The Hamiltonian of the system for a finite subset Λ of $\mathbf{Z}^\nu$ is

$$H_\Lambda = -\sum_{\substack{<j,k> \\ \in\Lambda}} J\,\sigma_j\,\sigma_k - B\sum_{j\in\Lambda}\sigma_j\,. \tag{83}$$

Here the symbol $<j,k>$ means that the sum is taken over all nearest neighbor pairs in Λ. J is a constant which is positive when one wishes to describe a model ferromagnet. (Then the energy arising from the first term in H_Λ is lower when a nearest neighbor pair of spins are aligned than when they are opposed.) The expectation value of a product of spins, $\prod_{i\in S}\sigma_i$, is given by

$$<\prod_{i \in S} \sigma_i>_\Lambda = [Z_\Lambda]^{-1} \sum_{\sigma_i;\ i \in \Lambda} \left[\prod_{j \in S} \sigma_j\right] \exp(-\beta H_\Lambda) \tag{84}$$

with

$$Z_\Lambda = \sum_{\sigma_i;\ i \in \Lambda} \exp(-\beta H_\Lambda) .$$

For $B = 0$, the Hamiltonian is invariant under the transformation taking each σ_i into its negative $-\sigma_i$, so that if, as in fact happens, at high temperatures there is a unique phase, one has for each Λ, $<\sigma_i>_\Lambda = 0$, and therefore in the thermodynamic limit, $<\sigma_i> = 0$. On the other hand, one would expect for $B > 0$, $<\sigma_i> > 0$, *since making the spins positive* lowers the energy by virtue of the second term. Lenz proposed the system as the simplest model in which ferromagnetism could arise as a cooperative phenomenon. Ferromagnetism for the model is defined as a discontinuity in $<\sigma_i>$ as a function of B at the value $B = 0$:

$$\lim_{B \to 0+} <\sigma_i> = -\lim_{B \to 0-} <\sigma_i> \neq 0 .$$

The value of this limiting quantity is called the *spontaneous magnetization.* This is a natural terminology if one regards the term $-B \sum_{i \in \Lambda} \sigma_i$ as an analogue of the energy of the spins in Λ in a magnetic field B and $<\sigma_i>$ as the *magnetization per unit volume.*

For the Lenz-Ising model on a one-dimensional lattice, $\nu = 1$, Ising proved in 1924 [62] that there is no ferromagnetism, and conjectured that the same should be true for lattices in the plane and in three dimensions. Peierls showed the contrary for the plane [63]: there is a temperature $T_c > 0$ such that for $T \leq T_c$ the spontaneous magnetization is non-vanishing. T_c is usually called the *Curie point.* Onsager gave an explicit formula for the free energy of the planar Lenz-Ising model with $B = 0$, above and below the Curie point [64]. This formula made it evident that most of the approximation methods which had been used up to that time work poorly in a neighborhood of the Curie point. See Figure 23(b). Onsager's exact solution played an important role historically because it made clear that the behavior of thermodynamic systems in a neighborhood of the critical point is a subtle problem.

From a knowledge of the behavior of an Ising ferromagnet, $J > 0$, one can get some information on the behavior of an Ising *anti-ferromagnet* by recognizing that if all the spins of the sublattice Λ_1 of Figure 22 are reversed, the Hamiltonian for a finite region is affected in two ways [65]. First, $J \to -J$ and second the interaction with the magnetic field, $-B(\sum_{i \in \Lambda} \sigma_i)$, is replaced by

$$-B\left[-\left(\sum_{i \in \Lambda_1 \cap \Lambda} \sigma_i\right) + \left(\sum_{i \in \Lambda \setminus (\Lambda_1 \cap \Lambda)} \sigma_i\right)\right]. \tag{85}$$

This latter interaction may be regarded as the interaction of the spins with a *staggered magnetic field.* The phase diagram of an anti-ferromagnet

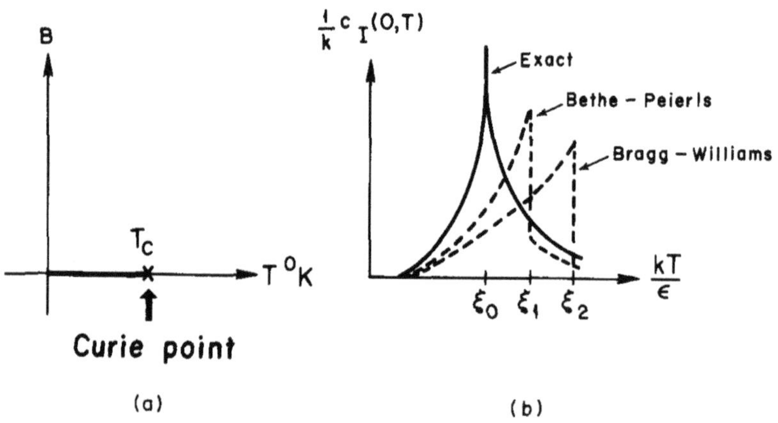

Fig. 23. a) The translation invariant, pure, equilibrium phases of the Lenz-Ising model for dimensions of the lattice ≥ 2; there is a cut from $T = 0$ to T_c and two distinct phases in this range of temperatures are obtained by approaching the cut from above or below. For all other values, $|B| \neq 0$ or $B = 0$, $T > T_c$ there is a unique phase.

b) Approximations to the specific heat for the plane Lenz-Ising model for $B = 0$ compared with Onsager's exact result [66].

is the same, Figure 23(a), as that of the feromagnet with the same value of $|J|$ provided B is interpreted as a staggered magnetic field.

For an anti-ferromagnet in the two phase region, the spontaneous magnetization does not distinguish the two phases; it is exactly zero. There is, however, a non-vanishing *sublattice magnetization*

$$\lim_{\Lambda \to \infty} < \frac{1}{|\Lambda|} \sum_{i \in \Lambda_1 \cap \Lambda} \sigma_i >_\Lambda$$

which takes an opposite sign in the two phases. There is a general lesson in this. The sublattice magnetization provides a reasonable *order parameter* for the phase transition to the anti-ferromagnetic phases. Namely, it vanishes in the single phase region and together with the temperature gives a parametrization of the pure phases in the two phase region. (These are the two indispensible characteristics of an order parameter according to the conventional wisdom originated by Landau.) However, as we have explained in connection with the definition of state in thermodynamics, the sublattice magnetization, is somewhat questionable as a macroscopic thermodynamic variable. Thus, we see that Lenz-Ising model reproduces the phenomena suggested by the thermodynamics. A generous definition of macroscopic variable will be necessary if equilibrium states are to be parametrized by macroscopic variables. If we ask whether the Lenz-Ising model or its more elaborate and realistic relative, the Heisenberg model actually are a good description of real ferromagnets, we have to answer, not often; the spins are carried around by electrons and only in a limiting case is it a good approximation to regard them as fixed at lattice sites. Thus, the main justification for the great effort that has gone into the study of lattice systems is theoretical: lattice systems provide a testing ground on which one can evaluate the adequacy of general concepts of statistical mechanics. Lattice systems are mathematically somewhat easier to handle than general systems of interacting particles and they provide plenty of interesting phase transitions.

The general theory of lattice spin systems, as it was developed in the late 1960's by Gallavotti, Miracle Sole, Ruelle, Robinson and Lanford, considered multispin interactions involving an arbitrarily large number of spins. The sum $-\sum_{\substack{<j,k> \\ j,k \in \Lambda}} J \sigma_j \sigma_k$, appearing in the Hamiltonian of the Lenz-Ising model, was replaced by

$$H_\Lambda = \sum_{X \subseteq \Lambda} \Phi(X) \sigma_X . \tag{86}$$

Here X is a subset of the sites lying in the finite subset Λ of the lattice Z^ν; $\sigma_X = \prod_{i \in X} \sigma_i$ is the product of the spins at those sites; $\Phi(X)$ is the interaction constant among the spins in X, the analogue of $-J$ for the nearest neighbor pairs. Evidently, an interaction, Φ, is then a real-valued function defined for every finite subset of the lattice. (By convention $\Phi(X) = 0$, for X the empty set.) Using this Hamiltonian as H_Λ, one can proceed first as in the Lenz-Ising model to define a thermodynamic function, p_Λ, which it is nowadays customary to call the pressure

$$p_\Lambda(\Phi) = |\Lambda|^{-1} \ell n \, Z_\Lambda(\Phi) \tag{87}$$

$$Z_\Lambda(\Phi) = \langle e^{-H_\Lambda} \rangle_0 . \tag{88}$$

Here

$$\langle A \rangle_0 = \prod_{i \in Z^\nu} \left(\frac{1}{2} \sum_{\sigma_i = \pm 1} \right) A(\sigma) ; \tag{89}$$

For a function A depending only on a finite number of spins, this is a finite sum. For general studies, the factor β, which would normally multiply H_Λ, is set equal to 1, or equivalently incorporated in Φ. The authors mentioned made the remarkable discovery that the thermodynamic limit of the pressure exists for all interactions satisfying

a) Translation Invariance

$$\Phi(X) = \Phi(X+a) \qquad a \in \mathbf{Z}^{\nu} \tag{90}$$

b) $\qquad |||\Phi||| < \infty \quad \text{or} \quad \|\Phi\|_{\sim} < \infty$

where

$$|||\Phi||| = \sum_{X \ni 0} \frac{|\Phi(X)|}{|X|}, \quad \|\Phi\|_{\sim} = \sum_{X \ni 0} |\Phi(X)| < \infty . \tag{92}$$

The set of Φ's satisfying these conditions form two separable real Banach spaces, $\mathcal{B}$ and $\tilde{\mathcal{B}}$ respectively. They further showed that the pressure satisfies the inequalities

$$|p(\Phi)| \leq |||\Phi||| \qquad |p(\Phi) - p(\Psi)| \leq |||\Phi - \Psi||| \tag{93}$$

and is a convex function of the interaction

$$p(\alpha\Phi + (1-\alpha)\Psi) \leq \alpha p(\Phi) + (1-\alpha)\, p(\Psi)$$

for $0 \leq \alpha \leq 1$. This begins to look like Gibbs' thermodynamics all over again except that it is in a new setting: the convex thermodynamic function p is now regarded as a function of the interaction, Φ. The parallel becomes even more striking if we inquire into the significance of the tangent planes to the graph of the pressure

$$dp_{\Lambda} = |\Lambda|^{-1} Z_{\Lambda}^{-1}\, dZ_{\Lambda} = Z_{\Lambda}^{-1} < \frac{-dH_{\Lambda}}{|\Lambda|} e^{-H_{\Lambda}} >_0 \tag{94}$$

where

$$dH_{\Lambda} = \sum_{X \subseteq \Lambda} d\Phi(X)\, \sigma_X .$$

Let $d\Phi(X) = \Psi(X)$. Then the linear form in the interactions which describes the tangent to the graph of the pressure at Φ is given by

$$dp_{\Lambda}(\Psi) = Z_{\Lambda}^{-1} < \frac{-H_{\Lambda}(\Psi)}{|\Lambda|} e^{-H_{\Lambda}} >_0 \tag{95}$$

which is a linear combination of correlation functions. If, in particular, we pick just one component of $\Psi \neq 0$, say $\Psi(X)$, we get the particular correlation function for the spin product σ_X, up to a factor

$$\frac{\Psi(X)}{|\Lambda|} M_{\Lambda}^{-}(X) \tag{96}$$

where M_{Λ}^{-} is the number of distinct subsets of Λ obtainable from X by translations of the lattice. (Do not forget that all interactions under discussion are translation invariant.) Thus, for the finite system in Λ, the state in the sense of statistical mechanics as determined by the correlation functions is given by the tangent to the pressure. Passage to the thermodynamic limit yields the corresponding statement for the state of the infinite system. Although the pressure for the finite system is strictly convex so the tangent plane at Φ is unique, the limiting pressure need not be and so we have the possibility of the various edges and corners in its graph discussed already in simple cases in connection with Figures 11 and 12.

The limiting form of the relation (95) connects a continuous linear functional $\alpha(\Psi)$ on the Banach space $\mathcal{B}$ to the expectation value of an observable A_{Ψ} in a state ρ

$$\alpha(\Psi) = -\rho(A_{\Psi}) \tag{97}$$

where

$$A_{\Psi} = \sum_{X \ni 0} \frac{\Psi(X)\,\sigma_X}{|X|}. \tag{98}$$

In the more general context considered by Israel it is the starting point, equation (2) of Chapter II, of his analysis of the connection between invariant states and tangents to the pressure. (The characteristic expression (98) appears because in an appropriate version of the thermodynamic limit $|\Lambda|^{-1} M_\Lambda^-(X)$ in (96) approaches 1 and the summation in H_Λ can be rearranged to yield $\sum_{X \ni 0}$ at the price of over counting X, $|X|$ times.) He proves the remarkable result that a linear functional, α, on $\mathfrak{B}$ is of the form (97) with ρ a state, invariant under translations of the lattice, if and only if α is p-bounded in the sense that there exists a constant C such that

$$p(\Psi) \geq \alpha(\Psi) + C \tag{99}$$

for all $\epsilon \mathfrak{B}$. Further he shows p-boundedness for α to be equivalent to finiteness of the entropy per unit volume of ρ. The linear functionals, α_Φ, that are tangent to the pressure at Φ, are certainly p-bounded since they satisfy

$$p(\Phi + \Psi) \geq p(\Phi) + \alpha_\Phi(\Psi)$$

and therefore

$$p(\psi) - \alpha_\Phi(\psi) \geq p(\Phi) - \alpha_\Phi(\Phi)$$

but as was first shown by Ruelle they are also characterized by equality in the variational principle (82). The equivalence of this characterization of invariant equilibrium states to the characterization by the DLR equilibrium equations (or for quantum statistical mechanics, the KMS conditions) is the subject of Israel's Chapter III.

The general framework discussed so far expresses the phase diagram of a lattice system in terms of the graph of the pressure as function of the interaction constants. To complete it, one needs a description of the decomposition of mixed phases into pure phases. This is the subject of Israel's Chapter IV based primarily on [60] and on [67]. There one finds in particular a proof that extremal invariant equilibrium phases are characterized by definite values of macroscopic observables, macroscopic observables being defined in this connection as spatial averages of local observables. This provides an answer to the question proved at the beginning of this introduction (how to define macroscopic variables to label equilibrium phases) but it is certainly not a complete answer, because it gives no information about the occurrence or non-occurrence of coexisting phases. As Gallavotti and Miracle-Sole already remarked such a phenomenon is exceptional in the sense of Mazur's theorem (interactions for which the tangent to the pressure is unique are a dense G_δ) but that does not give a very specific picture of the nature of those interactions which yield coexistence nor of the structure of the convex set of coexistence.

Here the methods of the last two chapters of Israel's thesis have yielded new results and opened up new possibilities for future work. Using his refinement of the Bishop-Phelps approximation theorem, Israel starts from an invariant state with some desired property and looks for an invariant equilibrium state with the same properties. He shows that the space has a dense set of interactions for which the equilibrium states show pathological behavior, but that in smaller spaces of interactions such as $\tilde{\mathcal{B}}$ such behavior does not occur. A concrete example of this kind of phenomenon is shown in Figure 24 [74].

On the other hand, in Chapter VI, it is shown that the set of points at which the structure of the set of coexisting phases is consistent with the Gibbs Phase Rule, at least in its Hausdorff dimension, is a dense G_δ.

The possibility of a more detailed description of the families of equilibrium phases is under study and further progress has been reported, both

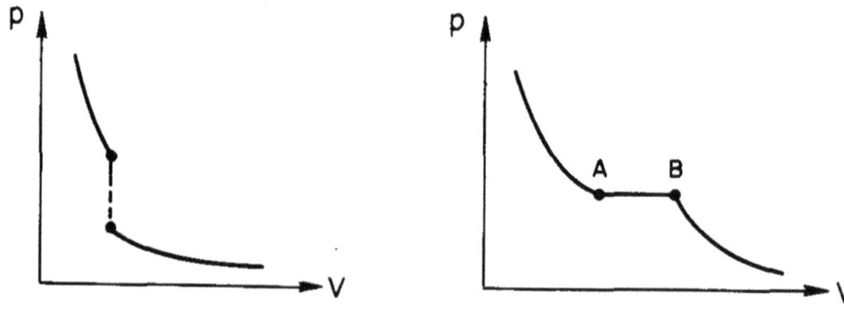

(a) (b)

Fig. 24. a) In the space $\mathcal{B}$ with a lattice gas interpretation, curves of constant temperature may be discontinuous as functions of the volume [68]. In $\tilde{\mathcal{B}}$ such behavior is impossible [69].

b) For $\tilde{\mathcal{B}}$ the isothermal is continuous. It is shown in a case with a coexistence region, AB.

in controlling the smoothness of the set of coexistence in general [70] and in parametrizing it explicitly in specific cases [71].

ACKNOWLEDGEMENT

I thank the Ecole Polytechnique for its hospitality during the completion of this introduction. I thank numerous colleagues for suggestions and corrections. The work was supported in part by NSF contract MPS 74-22844.

REFERENCES

[1] J. W. Gibbs, *Graphical Methods in the Thermodynamics of Fluids*. Transactions of the Connecticut Academy II (1873) 309-342 Coll. Works I 1-32.

[2] ————, *A Method of Geometrical Representation of the Thermodynamic Properties of Substances by Means of Surfaces*. Transactions of the Connecticut Academy II (1873) 382-404 Coll. Works 33-54.

[3] R. Clausius, *Über die bewegende Kraft der Wärme und die Gesetze, welche sich daraus fur die Wärmelehre selbst ableiten lassen.* Ostwald's Klassiker Nr. 99.

[4] See [1] footnote ‡ page 2. Gibbs assumes differentiability of U as a function of S and V. As will be seen in Figure 24(a), there are examples in statistical mechanics in which U is not everywhere differentiable. In most versions of thermodynamics only the special case treated by Gibbs is considered.

[5] See [2] p. 42 where he says of the graph of U as a function of V and S: "... if the form of the surface be such that it falls above the tangent plane except at the single point of contact, the equilibrium is necessarily stable..." and "... if the surface have such a form that any part of it falls below the fixed tangent plane, the equilibrium will be unstable." Incidentally, in Gibbs' terminology what we are here calling convex is concave upward. It has been pointed out to me by R. Jost that the thesis of M. Planck *(1881) also* emphasizes the physical significance of the concavity of the entropy in terms of thermodynamic stability.

[6] J. C. Maxwell *Theory of Heat* New Edition, D. Appleton, N. Y. (1875) *Representation of the Properties of a Substance by Means of a Surface* pp. 195-208.

[7] H. B. Callen, *Thermodynamics an introduction to the physical theories of equilibrium thermostatics and irreversible thermodynamics.* John Wiley, New York, 1960.

[8] L. Tisza, *Generalized Thermodynamics.* M.I.T. Press, Cambridge, Mass., 1966.

[9] The surface is described in Maxwell's *Theory of Heat.* A photograph appears in [11], and the surface itself is on display in the Sloane Physics Laboratory of Yale University.

[10] M. W. Zemansky and R. C. Herman, *The Gibbs and Mollier Thermodynamic Surfaces.* The American Physics Teacher *4* (1936) 194-196.

[11] A. L. Clark and L. Katz, *Thermodynamic Surfaces of* H_2O. Trans. Roy Soc. Canada (1939) 59-75.

[12] *Commentary on the works of J. Willard Gibbs*, ed. by E. G. Donnan and A. Haas, Yale University Press, 1932. •

[13] See [7], pp. 24-5. Callen says (p. 24): "The basic problem of thermodynamics is the determination of the equilibrium state that eventually results after the removal of internal constraints in a closed composite system."

[14] J. W. Gibbs, *On the Equilibrium of Heterogeneous Substances.* Transactions of the Connecticut Academy III (1875-6) 108-248 (1877-8) 343-524 Coll. Works I 55-353.

[15] For an argument starting from the droplet model see M. E. Fisher, *Theory of Condensation and the Critical Point.* Physics *3* (1962) 255-283. For a formal proof of the non-existence of metastable states (states understood in a certain technical sense) see O. E. Lanford III and D. Ruelle, *Observables at Infinity and States with Short Range Correlations.* Comm. in Math. Phys. *13* (1969) 124-215. For some reviews making good contact with experiment, see *Fluctuations, Instabilities, and Phase Transitions,* Ed. T. Riste, Plenum Press, New York, 1975.

[16] Needless to say both Gibbs and von der Waals were heavily influenced by the first experimental discovery of a critical point by T. Andrews. For further discussion of this point see E. H. Stanley, p. 46-49 in *Local Properties at Phase Transitions.* Proceeding of the International School of Physics "Enrico Fermi" Course LIX Ed. K. A. Muller. North Holland Pub. Co., Amsterdam, 1976.

[17] The argument uses that U is homogeneous of first degree in $S, V, N_1, \cdots, N_n$ and that therefore Euler's differential equation is

$$U = \left(\frac{\partial U}{\partial S}\right) S + \left(\frac{\partial U}{\partial V}\right) V + \sum_{j=1}^{n} \left(\frac{\partial U}{\partial N_j}\right) N_j$$

$$U = TS - pV + \mu_1 N_1 + \cdots \mu_n N_n$$

which, solved for $-p$ yields the stated relation.

[18] W. Fenchel, *On Conjugate Convex Functions.* Canadian Jour. Math. *1* (1949) 23-7.

[19] See [14], pp. 96-7.

[20] I thank B. Souillard for suggesting this construction.

[21] For reviews of van der Waal's various contributions and their significance in the light of our present knowledge see *Proceedings of the Van der Waals Centennial Conference on Statistical Mechanics,* Ed. C. Prins. Physica *73* (1973) 1-257.

[22] H. Kamerlingh Onnes Verslagen Kon. Akad. Wet. Amsterdam *21* (1881) Arch. Neerlandaises *30* (1897) 101.

[23] See [7], Chapter 14.

[24] For a review see E. G. D. Cohen, *Tricritical Points in Metamagnets and Helium Mixtures*, pp. 47-79 in Fundamental Problems in Statistical Mechanics III Proc. of the 1974 Wageningen Summer School Ed. E. G. D. Cohen. North Holland/American Elsevier Amsterdam, 1975.

[25] R. B. Griffiths and B. W. Widom, *Multi component Fluid Tricritical Points.* Phys. Rev. *A8* (1973) 2173-5.

[26] B. Widom, *Critical Phenomena,* pp. 1-45 in the summer school volume referred to in [24].

[27] R. B. Griffiths, *Thermodynamic Model for Tricritical Points in Ternary and Quaternary Fluid Mixtures.* Jour. Chem. Phys. *60* (1974) 195-206.

[28] J. M. H. Levelt Sengers, *From Van der Waals Equation to the Scaling Laws*, pp. 73-106 in [21].

[29] J. L. W. V. Jensen, *Sur les fonctions convexes et les inégalités entre les valeurs moyennes.* Acta Math. *30* (1906) 175-193.

[30] O. Stolz, *Grundzüge der Differential und Integralrechnung* I. Teubner, Leipzig, 1893.

[31] O. Hölder, *Uber einen Mittelwertsatz.* Gött. Nachrichten (1889) 38-41.

[32] D. Hilbert, *Gedächtnisrede auf H. Minkowski*, pp. V-XXXII in *H. Minkowski Gesammelte Abhandlungen.* Chelsea, New York, 1967.

[33] See pp. 131 ff. of his *Gesammelte Abhandlungen.*

[34] N. Bourbaki, *Eléments de Mathématiques* Livre V *Espaces Vectoriels Topologiques* Chapitre II *Ensembles Convexes et Espaces Localement Convexes.* 2ème Edit. Hermann, Paris, 1966.

[35] C. Caratheodory, *Untersuchungen uber die Grundlagen der Thermodynamik.* Math. Ann. *67* (1909) 355-386.

[36] See, for example, S. Karlin, *Mathematical Methods and Theory in Games, Programming, and Economics* I, II. Addison-Wesley, Reading, Mass., 1959.

[37] See, for example, R. B. Holmes, *Geometric Functional Analysis and its Applications.* Springer-Verlag, New York, 1975, p. 34.

[38] See [34], p. 82.

[39] R. R. Phelps, *Lectures on Choquet's Theorem.* van Nostrand, Princeton, N. J., 1966. See also O. E. Lanford, III *Selected Topics in Functional Analysis*, pp. 109-214 in *Méchanique statistique* et theorie quantique des Champs les Houches 1970. Eds. R. Stora and Ç. de Witt, Gordon and Breach, New York, 1971, especially pp. 123-140.

[40] S. Mazur, *Uber konvexe Menge in linearen normierten Raumen.* Studia Math *4* (1933) 70-84.

[41] O. E. Lanford and D. W. Robinson, *Statistical Mechanics of Quantum Spin Systems* III. Commun. Math. Phys. *9* (1968) 327-338.

[42] E. Bishop and R. R. Phelps, *The support functionals of a convex set.* Amer. Math. Soc. Symposia in Pure Math. Vol. 7, pp. 27-35, Amer. Math. Soc. Providence, Rhode Island, 1963.

[43] See [34], p. 155, Exercise 14.

[44] See [34], p. 107.

[45] J. Willard Gibbs, *Elementary Principles in Statistical Mechanics Developed With Especial Reference to the Rational Foundation of Thermodynamics.* Yale University Press, 1902.

[46] See [45], p. 166.

[47] G. Uhlenbeck (private communication) has suggested the phrase *bulk limit* as an alternative to thermodynamic limit, but in the following the more standard, if regrettable, terminology is employed.

[48] L. van Hove, *Quelques propriétés générales de l'intégrale de configuration d'un système de particules avec interaction.* Physica 15 (1949) 951-961.

[49] T. D. Lee and C. N. Yang, *Statistical Theory of Equations of State and Phase Transitions I Theory of Condensation.* Phys. Rev. *87* (1952) 404-419.

[50] For references to the developments of the 1960's see D. Ruelle, *Statistical Mechanics Rigorous Results.* W. A. Benjamin, New York, 1969.

[51] See [50], p. 36.

[52] See [45], p. 133.

[53] See [45], p. 167.

[54] See the papers of Ornstein and Zernike and of Zernike reprinted in *The Equilibrium Theory of Classical Fluids.* Ed. H. L. Frisch and J. L. Lebowitz. W. A. Benjamin, New York, 1964.

[55] N. Weiner. See J. E. Mayer and E. Montroll. Jour. Chem. Phys. *9* (1941) 2.

[56] See H. Van Beijeren. *Intersurface Sharpness in the Ising System.* Commun. in Math. Phys. *40* (1975) 1-6 and earlier references quoted there. The original proof is contained in R. L. Dobrushin, *Gibbs State Describing Coexistence of Phases for a Three-Dimensional Ising Model.* Theory Probability and Appl. *17* (1972) 582-600.

[57] J. Lebowitz, *Number of Phases in One Component Ferromagnets.* Jour. Stat. Phys., to appear.

[58] D. Ruelle [50], pp. 161-3.

[59] See [50], pp. 100-104 and Exercise 4E, p. 106.

[60] See [50], Chapter VI.

[61] D. Kastler, M. Mebkhout, G. Loupias, L. Michel, *Central Decomposition of Invariant States Applications to Groups of Time Translations and of Euclidean Transformations in Algebraic Field Theory.* Commun. Math. Phys. *27* (1972) 195-222.

[62] E. Ising, *Beitrag zur Theorie des Ferromagnetismus.* Zeits. für Phys. *31* (1925) 253-8.

[63] R. Peierls, *On Ising's Model of Ferromagnetism.* Proc. Comb. Phil. Soc. *32* (1936) 477-481. Peierls' argument was made rigorous by Griffiths. R. B. Griffiths-Peierls Proof of Spontaneous Magnetization in a Two-Dimensional Ising Ferromagnet. Phys. Rev. *136A* (1964) 437-439.

[64] L. Onsager, *Crystal Statistics* I, *A Two Dimensional Model with an Order Disorder Transition.* Phys. Rev. *65* (1944) 117-149.

[65] See [50], p. 143, Exercise 5.E.

[66] K. Huang, *Statistical Mechanics.* John Wiley, New York, 1963, Figure 17.6, p. 372.

[67] O. E. Lanford and D. Ruelle. *Observables at Infinity and Short Range Correlations in Statistical Mechanics.* Commun. in Math. Phys. *13* (1969) 194-215.

[68] M. Fisher, *On the Discontinuity of the Pressure.* Commun. in Math. Phys. *26* (1972) 6-14.

[69] R. C. Griffiths, D. Ruelle, *Strict convexity ("continuity") of the pressure in lattice systems.* Commun. in Math. Phys. *23* (1971) 194-215.

[70] D. Ruelle, *A Heuristic theory of Phase transitions.* Commun. Math. Phys. *53* (1977) 195-208.

[71] W. Holsztyński and J. Slawny, *Phases of Ferromagnetic Spin Systems at low temperatures.* Lettere al Nuovo Cim. *13* (1975) 534-8.

[72] An up-to-date review of the properties of entropy is A. Wehrl *General Properties of Entropy.* Rev. Mod. Phys. April 1978. The first proof of the concavity of the entropy for quantum systems and tempered interaction potentials is contained in R. B. Griffiths, *Microcanonical Ensemble in Quantum Statistical Mechanics.* Jour. Math. Phys. *6*

(1965) 1447-1461. It is extended to Coulomb potentials in *The Constitution of Matter: Existence of Thermodynamics for Systems Composed of Electrons and Nuclei.* Advances in Math. *9* (1972) 316-398.

[73] See J. L. Lebowitz and O. Penrose, *On the Exponential Decay of Correlations.* Commun. Math. Phys. *39* (1974) 165-184; and D. Iagolnitzer and B. Souillard, *Decay of correlations for slowly decreasing potentials.* Phys. Rev. *A16* (1977) 1700-1704. For slowly decreasing potentials, it is established in many cases that the correlation decreases as the potential does.

[74] It follows from convexity alone that $\frac{U}{V}\left(\frac{S}{V}, 1, \frac{N}{V}\right)$ has left and right derivatives with respect to N/V at each point. Since edges in the graph of U/V correspond to flat places in the graph of its conjugate convex function $p(T, \mu)$, Gibbs' assumption of differentiability of U/V is equivalent to strict convexity of $p(T, \mu)$. The discontinuities of the pressure as a function of the volume (or density) shown in Figure 24(a) arise from intervals of constancy of $\rho = \frac{\partial p}{\partial \mu}(T, \mu)$ as a function of μ.

Convexity
in the
Theory of Lattice Gases

I. INTERACTIONS

I.1. *Classical lattice systems*

The Ising model is a simple example of a statistical-mechanical system. At each site of the lattice $\mathbf{Z}^\nu$ we assume there is a "spin" which can be either "up" (+1) or "down" (−1). Thus for each subset Λ of $\mathbf{Z}^\nu$ we have the space $\Omega_\Lambda = \{-1,+1\}^\Lambda$ of configurations in Λ. With the product topology this forms a compact metric space, even if Λ is infinite. The full configuration space $\Omega_{\mathbf{Z}^\nu}$ will be denoted simply by Ω, and $\sigma_i(\omega)$ will denote the spin at site $i \in \mathbf{Z}^\nu$ in the configuration ω. The Hamiltonian for the system in a finite subset Λ of $\mathbf{Z}^\nu$ will be some real-valued function H_Λ on Ω_Λ. At inverse temperature β we obtain the partition function (in a canonical ensemble)

$$Z_\Lambda(\beta) = \sum_{\omega \in \Omega_\Lambda} e^{-\beta H_\Lambda(\omega)} \tag{1}$$

and thermal averages

$$\langle A \rangle_{\beta,\Lambda} = Z_\Lambda(\beta)^{-1} \sum_{\omega \in \Omega_\Lambda} A(\omega) e^{-\beta H_\Lambda(\omega)} \tag{2}$$

for functions A on Ω_Λ.

For more general classical lattice systems, we will allow the "spin" at each site to vary over a compact metric space Ω_0. The configuration space corresponding to $\Lambda \subset \mathbf{Z}^\nu$ is then $\Omega_\Lambda = (\Omega_0)^\Lambda$, again with the product topology. Again $\Omega_{\mathbf{Z}^\nu}$ will be denoted by Ω. The Hamiltonian for a finite subset Λ of $\mathbf{Z}^\nu$ will be a continuous real-valued function H_Λ on Ω_Λ. To obtain a partition function and thermal averages we need an *"a priori"* measure to take the place of the summations in (1) and (2). It will be convenient to use a probability measure for this purpose. This measure should describe the state of a noninteracting system, and so the spins at different sites should be independent and identically distributed under this measure. Thus we will take a probability measure μ_0 on Ω_0, and the *a priori* measure for $\Lambda \subset Z^\nu$ will be the product measure μ_0^Λ on Ω_Λ; expectation values under μ_0^Λ will frequently be denoted $<F>_{0,\Lambda}$. Whenever there is no danger of confusion we will simply use μ_0 and $<F>_0$ in place of μ_0^Λ and $<F>_{0,\Lambda}$. We also use the natural maps $i_{XY}: \Omega_X \to \Omega_Y$ (for $Y \subset X \subset Z^\nu$) without warning; thus we will tacitly identify a function F on Ω_Y with the function $F \circ i_{XY}$ on Ω_X. We will always assume that μ_0 is supported on all of Ω_0, i.e. every non-empty open subset has nonzero μ_0-measure.

In many cases the choice of μ_0 will be suggested by some symmetry of Ω_0. For example, if Ω_0 is a finite set we will take μ_0 to be normalized counting measure; in the classical Heisenberg model Ω_0 is a sphere, so we will take μ_0 to be normalized surface measure. In other cases the choice will not be so obvious; we venture no opinion on how Nature chooses an *a priori* measure.

The partition function and thermal averages for our system in a finite subset Λ of Z^ν are now

(3) $$Z_\Lambda(\beta) = <e^{-\beta H_\Lambda}>_0$$

and

$$<A>_{\beta,\Lambda} = Z_\Lambda(\beta)^{-1} <A e^{-\beta H_\Lambda}>_0 .$$

Note that these are analytic functions of β and of any parameter entering linearly in the Hamiltonian. Thus the finite system has no phase transitions; to find such interesting phenomena we must pass to an infinite-volume limit. From the physical point of view this limit should provide a description of the properties of the system "in bulk," removing all surface effects.

Our Hamiltonians will arise from interactions. For the Ising model an interaction is usually taken as a real function ϕ on finite nonempty subsets of Z^ν, and the associated Hamiltonians are

$$H_\Lambda^\phi = \sum_{X \subset \Lambda} \phi(X)\sigma_X \quad \text{where } \sigma_X(\omega) = \prod_{i \in X} \sigma_i(\omega) . \tag{5}$$

Now instead of speaking of spins we could use a different formulation of the same model: the "lattice gas." Here the configuration space for a subset Λ of Z^ν is taken to be the set $\mathcal{P}(\Lambda)$ of subsets of Λ. This can be identified with Ω_Λ by $\omega \longleftrightarrow \{i \in \Lambda: \sigma_i(\omega) = +1\}$. In the lattice gas formulation the interaction is again a real function $\tilde{\phi}$ on finite nonempty subsets of Z^ν, but the Hamiltonian in a finite subset Λ of Z^ν is the function on $\mathcal{P}(\Lambda)$ given by

$$H_\Lambda^{\tilde{\phi}}(X) = \sum_{Y \subset X} \tilde{\phi}(Y) . \tag{6}$$

Under the identification of $\mathcal{P}(\Lambda)$ with Ω_Λ this becomes the function on Ω_Λ given by

$$H_\Lambda^{\tilde{\phi}} = \sum_{Y \subset \Lambda} \tilde{\phi}(Y)\rho_Y \tag{7}$$

where $\rho_Y(\omega) = \begin{cases} 1 & \text{if } \sigma_i(\omega) = +1 \text{ for all } i \in Y \\ 0 & \text{otherwise.} \end{cases}$

We will use a more general formulation which includes both spin and lattice gas interactions as special cases. This will have several advantages:

i) it can be applied to an arbitrary Ω_0 and μ_0, where there are no obvious choices to replace the σ_X or ρ_X above;

ii) it provides a simple way of expressing certain interactions we will use in Chapters II and III;

iii) it provides a space of interactions with a norm naturally connected to the norm topology on states (probability measures on the configuration space).

In our formulation, an *interaction* Φ assigns to each nonempty finite subset X of Z^ν a real-valued continuous function $\Phi(X)$ on Ω_X. We will always assume our interactions are *translation-invariant*, i.e., they satisfy

$$\Phi(X+i) = \tau_i \Phi(X) \text{ for all } X \subset Z^\nu \text{ finite and nonempty, } i \in Z^\nu$$

where τ_i is the natural map from the space $C(\Omega_X)$ of continuous functions on Ω_X to $C(\Omega_{X+i})$. The Hamiltonian for a finite subset Λ of Z^ν is then the function H^Φ_Λ on Ω_Λ given by

$$H^\Phi_\Lambda = \sum_{X \subset \Lambda} \Phi(X) \tag{8}$$

In the Ising model, with $\Omega_0 = \{-1,+1\}$, a "spin language" interaction associated to a Hamiltonian as in (5) is identified with one of our interactions having each $\Phi(X)$ a real multiple of σ_X, while for a "lattice gas language" interaction as in (7) we would take each $\Phi(X)$ to be a multiple of ρ_X.

An interaction Φ has *finite range* if $\Phi(X) = 0$ for all X of sufficiently large diameter. The linear space of finite-range interactions will be denoted $\mathcal{B}_0$. Banach spaces of interactions are obtained by completing $\mathcal{B}_0$ in various norms. We denote by $\mathcal{B}$ the space of interactions with

$$|||\Phi||| \equiv \sum_{X \ni 0} \frac{\|\Phi(X)\|_\infty}{|X|} < \infty$$

where $|X|$ is the number of sites in X. $\tilde{\mathcal{B}}$ will denote the space of interactions with

$$\|\Phi\|_{\sim} \equiv \sum_{X \ni 0} \|\Phi(X)\|_\infty < \infty$$

These are separable real Banach spaces with $\mathcal{B}_0 \subset \tilde{\mathcal{B}} \subset \mathcal{B}$, $\mathcal{B}_0$ being dense in both $\tilde{\mathcal{B}}$ and $\mathcal{B}$.

For $A \in C(\Omega_X)$ real-valued, we define an interaction Ψ_A^X by

(9)
$$\Psi_A^X(X+i) = \tau_i A \quad \text{for } i \in Z^\nu$$
$$\Psi_A^X(Y) = 0 \qquad \text{if } Y \text{ is not a translate of } X.$$

These will form a particularly important class of interactions.

I.2. *The pressure**

We will call the function

(10)
$$P_\Lambda(\Phi) = |\Lambda|^{-1} \ln < e^{-H_\Lambda^\Phi} >_0$$

the pressure for the finite region $\Lambda \subset Z^\nu$. This terminology is really appropriate for the lattice gas, where the ensemble considered is grand canonical; in the canonical ensemble it would be more correct to speak of "minus the free energy per site," but we will ignore this distinction. Some of the importance of P_Λ arises from what its directional derivatives tell us about the thermal averages of observables. Suppose $A \in C(\Omega_X)$ is real-valued, and let Ψ_A^X be the interaction defined as in (9). Then since

$$H_\Lambda^{\Phi+t\Psi_A^X} = H_\Lambda^\Phi + t \sum_{i+X \subset \Lambda} \tau_i A$$

(11)
$$-\frac{d}{dt} P_\Lambda(\Phi + t\Psi_A^X)_{|t=0} = < e^{-H_\Lambda^\Phi} >_0^{-1} |\Lambda|^{-1} < \sum_{i+X \subset \Lambda} (\tau_i A) e^{-H_\Lambda^\Phi} >_0$$

*Sections I.2 and I.3 are largely adapted from [25], Chapter 2, and [28].

which is the thermal average of $|\Lambda|^{-1} \sum_{i+X \subset \Lambda} \tau_i A$ at inverse temperature $\beta = 1$ for the Hamiltonian H_Λ^Φ. Since β can be absorbed into the interaction, we will generally take $\beta = 1$ for simplicity. If we were only interested in a finite system, we could take $X = \Lambda$ and obtain the thermal average of A itself; however, we are interested in the infinite system, where a purely local perturbation (leaving out all the other translates of A) would have negligible effect. Our hope is that as Λ increases we *obtain the thermal average of* A in a translation-invariant state of the infinite system.

Our first task is to study the infinite-volume limit of $P_\Lambda(\Phi)$. After some preliminary inequalities, we will consider a sequence of (ν-dimensional) cubes C_a, and then show that the same limit is obtained for any sequence Λ_n tending to infinity in the sense of van Hove (defined below), which means roughly that the boundary of Λ becomes negligible in comparison to Λ itself.

LEMMA I.2.1 (Jensen's Inequality). *For any probability measure* μ *and real random variable* $f \in L^1(\mu)$,

$$e^{\int f\, d\mu} \leq \int e^f\, d\mu . \tag{12}$$

Proof. We can assume f is bounded below, since both sides of (12) are the limits as $N \to \infty$ of the expressions obtained by substituting $\max(f, -N)$ for f. Since adding a constant to f multiplies both sides by a positive constant, we can assume $f \geq 0$. Now by Hölder's Inequality $\int f\, d\mu \leq (\int f^n\, d\mu)^{1/n}$, so

$$e^{\int f d\mu} = \sum_{n=0}^{\infty} \frac{(\int f\, d\mu)^n}{n!} \leq \sum_{n=0}^{\infty} \frac{\int f^n\, d\mu}{n!} = \int e^f\, d\mu . \qquad \blacksquare$$

LEMMA I.2.2. *For any probability measure* μ *and real functions* $f,g \in L^\infty(\mu)$

$$\left| \ln \int e^f \, d\mu \right| \le \|f\|_\infty \tag{13}$$

$$\left| \ln \int e^f \, d\mu - \ln \int e^g \, d\mu \right| \le \|f-g\|_\infty \, . \tag{14}$$

Proof. Since $e^{-\|f\|_\infty} \le e^{-f} \le e^{\|f\|_\infty}$ almost everywhere,

$$e^{-\|f\|_\infty} \le \int e^{-f} \, d\mu \le e^{\|f\|_\infty} \, .$$

Taking logarithms yields (13). Since $d\mu_g = (\int e^g d\mu)^{-1} e^g \, d\mu$ is also a probability measure, Jensen's Inequality yields

$$\ln \int e^f \, d\mu - \ln \int e^g \, d\mu = \ln \int e^{f-g} \, d\mu_g \ge \int (f-g) \, d\mu_g \ge -\|f-g\|_\infty \, .$$

By interchanging f and g we also obtain

$$\ln \int e^f \, d\mu - \ln \int e^g \, d\mu \le \|f-g\|_\infty \, . \qquad \blacksquare$$

We can now obtain some useful estimates on P_Λ. Note that

$$\|H_\Lambda^\Phi\|_\infty \le \sum_{X \subset \Lambda} \|\Phi(X)\|_\infty = \sum_{i \in \Lambda} \sum_{\substack{X \subset \Lambda \\ i \in X}} \frac{\|\Phi(X)\|_\infty}{|X|} \le |\Lambda| \; |||\Phi||| \, . \tag{15}$$

Thus by Lemma I.2.2 we have

(16) $$|P_\Lambda(\Phi)| \le |\Lambda|^{-1} \|H_\Lambda^\Phi\|_\infty \le |||\Phi|||$$

(17) $$|P_\Lambda(\Phi) - P_\Lambda(\Psi)| \le |\Lambda|^{-1} \|H_\Lambda^\Phi - H_\Lambda^\Psi\|_\infty \le |||\Phi - \Psi||| .$$

Therefore if we can show that $P_\Lambda(\Phi)$ tends to a limit when $\Phi \in \mathcal{B}_0$, the usual type of "3ε" argument will extend this to $\mathcal{B}$.

THEOREM I.2.3. *For positive integers* a, *let* C_a *be a cube of side* a. *Then for* $\Phi \in \mathcal{B}$, $P(\Phi) = \lim_{a \to \infty} P_{C_a}(\Phi)$ *exists.* P *is a convex function on* $\mathcal{B}$ *with* $|P(\Phi) - P(\Psi)| \le |||\Phi - \Psi|||$.

Proof. As remarked above, we need only to prove this in $\mathcal{B}_0$ and then use (17) to extend it to $\mathcal{B}$. Let $\Phi \in \mathcal{B}_0$ and let d be a positive integer such that $\Phi(X) = 0$ whenever X has diameter at least d. Given a, let $b = n(d+a) + c$ with $n \ge 1$, $0 \le c < d+a$. We partition the cube C_b as shown in Figure 1A: the unshaded region Λ' is formed by n^ν cubes of side a, separated by corridors of width d. Since these cubes of side a do not interact with each other, $H_{\Lambda'}^\Phi$ is the sum of n^ν independent copies of $H_{C_a}^\Phi$ (functions on the configurations in disjoint cubes, and thus independent with respect to μ_0). Therefore

$$P_{\Lambda'}(\Phi) = n^{-\nu} a^{-\nu} \ln \langle e^{-H_{\Lambda'}^\Phi} \rangle_0 = a^{-\nu} \ln \langle e^{-H_{C_a}^\Phi} \rangle_0 = P_{C_a}(\Phi) .$$

Now

$$\left| |\Lambda'| P_{\Lambda'}(\Phi) - |C_b| P_{C_b}(\Phi) \right| = \left| \ln \langle e^{-H_{\Lambda'}^\Phi} \rangle_0 - \ln \langle e^{-H_{C_b}^\Phi} \rangle_0 \right|$$

$$\le \|H_{\Lambda'}^\Phi - H_{C_b}^\Phi\|_\infty \le \sum_{\substack{X \subset C_b \\ X \not\subset \Lambda'}} \|\Phi(X)\|_\infty \le |C_b \setminus \Lambda'| \, \|\Phi\| .$$

Dividing through by $|C_b| = b^\nu$ we have

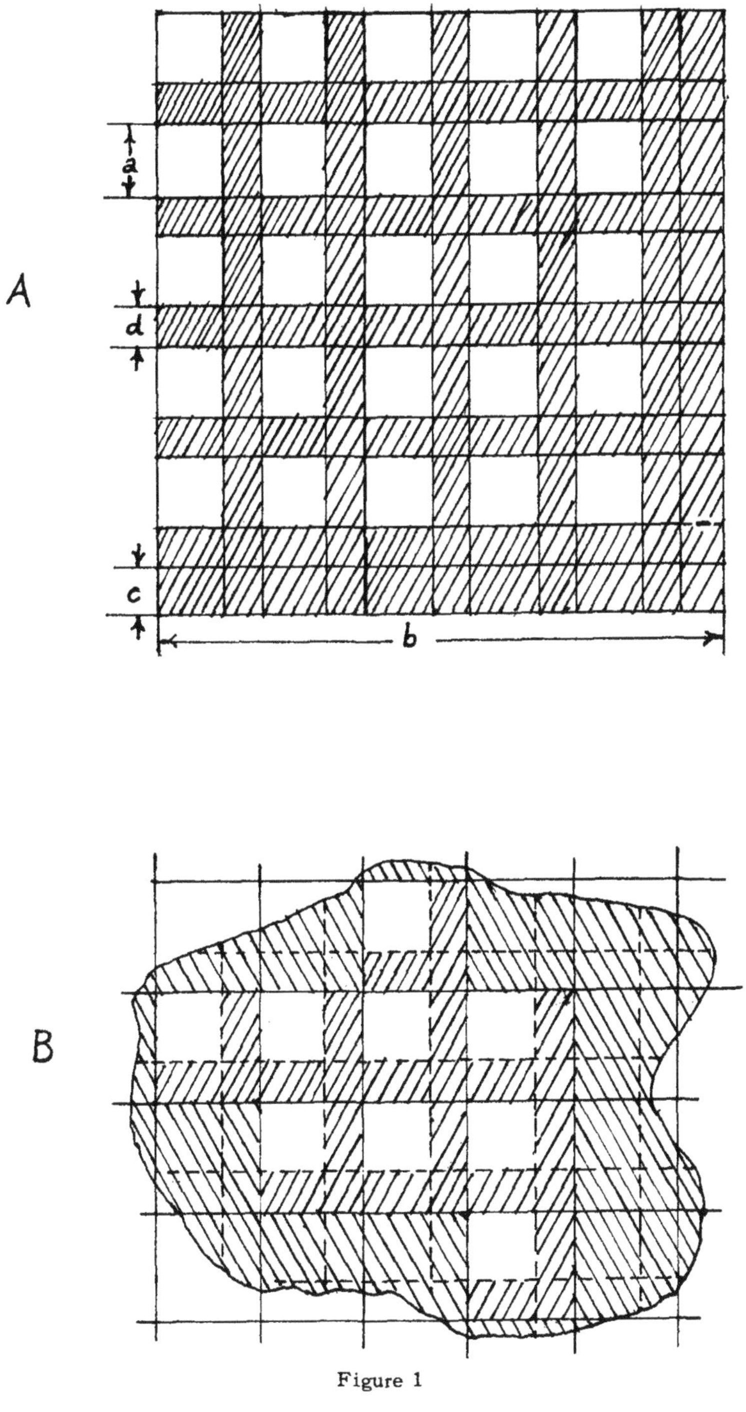

Figure 1

$$\left|\left(\frac{na}{b}\right)^{\nu} P_{C_a}(\Phi) - P_{C_b}(\Phi)\right| \leq \left(1 - \left(\frac{na}{b}\right)^{\nu}\right) \|\Phi\|_{\sim} .$$

As $b \to \infty$, $\frac{n(a+d)}{b} \to 1$, and so for each a

$$\limsup_{b,b' \to \infty} \left|P_{C_b}(\Phi) - P_{C_{b'}}(\Phi)\right| \leq 2\left(1 - \left(\frac{a}{a+d}\right)^{\nu}\right) \|\Phi\|_{\sim} .$$

Now taking $a \to \infty$, the $P_{C_b}(\Phi)$ *form a Cauchy sequence with limit* $P(\Phi)$. The estimate (17) carries over to the limit to yield

$$|P(\Phi) - P(\Psi)| \leq |||\Phi - \Psi||| .$$

Convexity of P_Λ is obtained by Hölder's Inequality:

$$< e^{-H_\Lambda^{t\Phi+(1-t)\Psi}} >_0 \; = \; < e^{-tH_\Lambda^{\Phi}} e^{-(1-t)H_\Lambda^{\Psi}} >_0 \; \leq \; < e^{-H_\Lambda^{\Phi}} >_0^{t} < e^{-H_\Lambda^{\Psi}} >_0^{1-t}$$

for $0 \leq t \leq 1$, and so

$$P_\Lambda(t\Phi + (1-t)\Psi) \leq tP_\Lambda(\Phi) + (1-t)P_\Lambda(\Psi) .$$

This inequality also carries over to the limit, yielding convexity of P. ■

For each positive integer a we partition Z^ν into a family $\mathcal{C}_a$ of cubes of the form $\{i: n_j a \leq i_j < (n_j+1)a,\ j = 1,\cdots,\nu\}$ for integers $n_1,\cdots,n_\nu$. For each finite subset Λ of Z^ν we define $N_a^+(\Lambda)$ as the number of cubes in $\mathcal{C}_a$ which intersect Λ, and $N_a^-(\Lambda)$ as the number of cubes in $\mathcal{C}_a$ contained in Λ. A sequence of finite subsets Λ_n of Z^ν will be said to *converge to infinity in the sense of van Hove* if for all a, $N_a^-(\Lambda_n) \to \infty$ and $N_a^-(\Lambda_n)/N_a^+(\Lambda_n) \to 1$.

THEOREM I.2.4. *If* $\Lambda_n \to \infty$ *(van Hove) and* $\Phi \in \mathcal{B}$, *then* $P_{\Lambda_n}(\Phi) \to P(\Phi)$.

Proof. By the usual "3ε" argument it suffices to prove this for $\Phi \in \mathcal{B}_0$.

We again take d so that $\Phi(X) = 0$ when $\mathrm{diam}(X) \geq d$. Fix a and n. There are $N^{-}_{a+d}(\Lambda_n)$ cubes in $\mathcal{C}_{a+d}$ contained in Λ_n, and by removing corridors of width d we are left with the unshaded region Λ' of Figure 1B, consisting of $N^{-}_{a+d}(\Lambda_n)$ cubes of side a. As in the proof of Theorem I.2.3 we have $P_{\Lambda'}(\Phi) = P_{C_a}(\Phi)$ and

$$\left| |\Lambda'| P_{\Lambda'}(\Phi) - |\Lambda_n| P_{\Lambda_n}(\Phi) \right| \leq \left\| H^{\Phi}_{\Lambda'} - H^{\Phi}_{\Lambda_n} \right\|_{\infty} \leq |\Lambda_n \setminus \Lambda'| \, \|\Phi\|_{\sim}$$

$$|P_{\Lambda_n}(\Phi) - |\Lambda_n|^{-1} a^{\nu} N^{-}_{a+d}(\Lambda_n) P_{C_a}(\Phi)| \leq (1 - |\Lambda_n|^{-1} a^{\nu} N^{-}_{a+d}(\Lambda_n)) \|\Phi\|_{\sim} .$$

As $n \to \infty$, $|\Lambda_n|^{-1} a^{\nu} N^{-}_{a+d}(\Lambda_n) \to \left(\frac{a}{a+d}\right)^{\nu}$, and so

$$\limsup \left| P_{\Lambda_n}(\Phi) - \left(\frac{a}{a+d}\right)^{\nu} P_{C_a}(\Phi) \right| \leq \left(1 - \left(\frac{a}{a+d}\right)^{\nu}\right) \|\Phi\|_{\sim} .$$

Now taking $a \to \infty$, we have $P_{C_a}(\Phi) \to P(\Phi)$ and $\left(\frac{a}{a+d}\right)^{\nu} \to 1$, so

$$P_{\Lambda_n}(\Phi) \to P(\Phi) . \qquad \blacksquare$$

So far we have been using "free boundary conditions": the Hamiltonian H^{Φ}_{Λ} describes a situation where there is nothing outside Λ to interact with the spins in Λ. For interactions in the space $\tilde{\mathcal{B}}$ we may also consider other types of boundary conditions. We will find that the infinite-volume pressure $P(\Phi)$ is independent of which boundary conditions are used. There are two main types of boundary conditions to consider: external fields (which might more accurately be called "external configurations") and periodic boundary conditions.

An *external configuration* for the finite subset Λ of Z^{ν} is a member τ of Ω_{Λ^c} (Λ^c denotes the complement of Λ in Z^{ν}). The Hamiltonian corresponding to the external configuration $\tau \in \Omega_{\Lambda^c}$ is the function on Ω_{Λ}

(18) $$ {}_{\tau}H_{\Lambda}^{\Phi}(\omega) = \sum_{X \cap \Lambda \neq \phi} \Phi(X)(\omega \times \tau) . $$

This sum converges if $\Phi \in \tilde{\mathcal{B}}$, and $\|{}_{\tau}H_{\Lambda}^{\Phi} - {}_{\tau}H_{\Lambda}^{\Psi}\|_{\infty} \leq |\Lambda| \, \|\Phi - \Psi\|_{\sim}$. The corresponding pressure $|\Lambda|^{-1} \ln \langle e^{-{}_{\tau}H_{\Lambda}^{\Phi}} \rangle_0$ will be denoted ${}_{\tau}P_{\Lambda}(\Phi)$.

For *periodic boundary conditions* we will consider the "torus" $T_n = (Z/nZ)^{\nu}$ with the quotient map $q_n \colon Z^{\nu} \to T_n$. The configuration space for T_n is $\Omega_{T_n} = (\Omega_0)^{T_n}$, and q_n induces a map $r_n \colon \Omega_{T_n} \to \Omega$ (repeating the configuration across Z^{ν} with period n in each coordinate direction). The Hamiltonian for T_n should be a sum of functions $\Phi(X) \circ r_n$ on Ω_{T_n}. We will want one term for each equivalence class $[X] = \{X + nj \colon j \in Z^{\nu}\}$. In order to facilitate the treatment of the quantum case, we restrict the sum to those X on which q_n is one-to-one, i.e., X contains no two points separated by a member of nZ^{ν}. The sum over equivalence classes $[X]$ for finite subsets X of Z^{ν} with this restriction will be denoted $\sum^*_{[X]}$. Thus the periodic-boundary-condition Hamiltonian is (the function on Ω_{T_n})

(19) $$ H_{T_n}^{\Phi} = \sum_{[X]}^{*} \Phi(X) \circ r_n . $$

Again this converges for $\Phi \in \tilde{\mathcal{B}}$, with $\|H_{T_n}^{\Phi}\|_{\infty} \leq n^{\nu} \|\Phi\|_{\sim}$. The corresponding pressure is $P_{T_n}(\Phi) = n^{-\nu} \ln \langle \exp(-H_{T_n}^{\Phi}) \rangle_0$.

THEOREM I.2.5. *Let* $\Phi \in \tilde{\mathcal{B}}$. *Then* $P_{T_n}(\Phi) \to P(\Phi)$, *and for any sequence* $\Lambda_n \to \infty$ *(van Hove) with external configurations* $\tau_n \in \Omega_{\Lambda_n^c}$, ${}_{\tau_n}P_{\Lambda_n}(\Phi) \to P(\Phi)$.

Proof. From the estimates on ${}_{\tau_n}H_{\Lambda_n}^{\Phi}$ and $H_{T_n}^{\Phi}$ we obtain

$$\left|{}_{\tau_n}P_{\Lambda_n}(\Phi) - {}_{\tau_n}P_{\Lambda_n}(\Psi)\right| \leq \|\Phi - \Psi\|_{\sim}$$

and

$$\left|P_{T_n}(\Phi) - P_{T_n}(\Psi)\right| \leq \|\Phi - \Psi\|_{\sim}.$$

Therefore it is enough to prove the theorem for $\Phi \in \mathcal{B}_0$, and then use a "3ε" argument to extend it to $\tilde{\mathcal{B}}$. Suppose $\Phi \in \mathcal{B}_0$, with $\Phi(X) = 0$ when $\operatorname{diam}(X) \geq d$. Let a be fixed. For the external configuration boundary conditions, we use a modification of the construction for Theorem I.2.4 to obtain $\Lambda' \subset \Lambda_n$ consisting of $N^-_{a+2d}(\Lambda_n)$ cubes of side a separated from each other and from Λ_n^c by a distance at least d. We then have

$$\left| |\Lambda_n| \, {}_{\tau_n}P_{\Lambda_n}(\Phi) - a^\nu N^-_{a+2d}(\Lambda_n) P_{C_a}(\Phi)\right| \leq \left\| {}_{\tau_n}H^\Phi_{\Lambda_n} - H^\Phi_{\Lambda'} \right\|_\infty$$

$$= \left\| \sum_{X \cap (\Lambda_n \setminus \Lambda') \neq \emptyset} \Phi(X)(\cdot \times \tau_n) \right\|_\infty \leq |\Lambda_n \setminus \Lambda'| \, \|\Phi\|_{\sim}.$$

Taking $n \to \infty$ and then $a \to \infty$ as in the proof of Theorem I.2.4, we obtain ${}_{\tau_n}P_{\Lambda_n}(\Phi) \to P(\Phi)$.

For periodic boundary conditions, let C' be a cube of side $n-d$ contained in T_n. Then we have

$$\left| n^\nu P_{T_n}(\Phi) - (n-d)^\nu P_{C_{n-d}}(\Phi)\right| \leq \left\| H^\Phi_{T_n} - H^\Phi_{C'} \right\|_\infty = \left\| \sum_{[X]:\, q_n(X) \not\subset C'} \Phi(X) \circ r_n \right\|_\infty$$

$$\leq |T_n \setminus C'| \, \|\Phi\|_{\sim}.$$

As $n \to \infty$ we obtain $P_{T_n}(\Phi) \to P(\Phi)$. ∎

I.3. *Quantum lattice systems*

For our purposes, the word "quantum" will signify the presence of noncommuting observables. In particular, since we are concerned with "spins" on a lattice rather than with particles on a continuum, there is no need to worry about Bose or Fermi statistics. At each site $i \in Z^\nu$ we assume there is a copy $\mathcal{H}_i$ of a finite-dimensional Hilbert space $\mathcal{H}_0$. For a "spin-½" system this would be two-dimensional; one orthonormal basis would consist of one vector representing an "up" spin and another representing a "down" spin. For each finite subset X of Z^ν we have the Hilbert space $\mathcal{H}_X = \bigotimes_{i \in X} \mathcal{H}_i$; the Hamiltonian for X would be some self-adjoint linear operator on $\mathcal{H}_X$.

To deal with the infinite system we will not try to construct an infinite tensor product of Hilbert spaces, which would be a rather intractable object; rather we will consider a *quasi-local* C^**-algebra* of observables. For each finite subset X of Z^ν the "local C^*-algebra" $\mathfrak{A}_X$ consists of all linear operators on $\mathcal{H}_X$. There are canonical injections $j_{XY}: \mathfrak{A}_Y \to \mathfrak{A}_X$ for $Y \subset X$, taking $A \in \mathfrak{A}_Y$ to the operator $A \otimes 1$ on $\mathcal{H}_X = \mathcal{H}_Y \otimes \mathcal{H}_{X \setminus Y}$. We can take an inductive limit of this system of C^*-algebras and injections, i.e., there is a C^*-algebra $\mathfrak{A}$ with injections $j_X: \mathfrak{A}_X \to \mathfrak{A}$, such that $j_X \circ j_{XY} = j_Y$ and $\bigcup_X j_X(\mathfrak{A}_X)$ is dense in $\mathfrak{A}$. We will henceforth leave out the j_{XY}'s and j_X's as much as possible, regarding the $\mathfrak{A}_X$ as subalgebras of $\mathfrak{A}$. For an infinite subset Λ of Z^ν we can define $\mathfrak{A}_\Lambda$ as the closure of the union of $\mathfrak{A}_X$ for all finite subsets X of Λ. It is easily seen that any member of $\mathfrak{A}_\Lambda$ commutes with any member of $\mathfrak{A}_{\Lambda^c}$. Translation in the lattice Z^ν induces a natural action of Z^ν on $\mathfrak{A}$ by automorphisms τ_i, taking $\mathfrak{A}_X$ to $\mathfrak{A}_{X+i}$.

A similar inductive-limit construction can be applied to the C^*-algebras $C(\Omega_X)$ of the classical lattice system, yielding $C(\Omega)$. Very similar roles will be played by $C(\Omega)$ and $\mathfrak{A}$ in the theories of classical and quantum systems respectively. Quasi-local C^*-algebras can be considered in somewhat greater generality (see [25], p. 148), but this will not be needed for our purposes. For example, we will make no attempt to consider an infinite-dimensional $\mathcal{H}_0$, which would be analogous to a nondiscrete Ω_0.

For a system in a finite region Λ with Hamiltonian H_Λ, the partition function and thermal averages of observables $A \in \mathfrak{A}_\Lambda$ (at $\beta = 1$ as usual) are

$$(20) \qquad \begin{aligned} Z_\Lambda &= \mathrm{tr}_\Lambda\, e^{-H_\Lambda} \\ \langle A \rangle_{1,\Lambda} &= Z^{-1}\, \mathrm{tr}_\Lambda(A\, e^{-H_\Lambda})\,. \end{aligned}$$

Here we are using the *normalized trace* $\mathrm{tr}_\Lambda(A) = d^{-1} \sum_{n=1}^{d} \langle u_n, Au_n \rangle$ where $\{u_1, \cdots, u_d\}$ is an orthonormal basis of $\mathcal{H}_\Lambda$. This will be more convenient than the more usual trace $\mathrm{Tr}_\Lambda(A) = \sum_{n=1}^{d} \langle u_n, Au_n \rangle$ since $\mathrm{tr}_X(j_{XY} A) = \mathrm{tr}_Y A$ for $Y \subset X$ and $A \in \mathfrak{A}_Y$. Moreover we have $|\mathrm{tr}_\Lambda A| \leq \|A\|$. Thus the normalized traces tr_Λ extend by continuity to a positive norm-one linear functional (a *state*) on $\mathfrak{A}$, which we denote tr. This will play a similar role to that of the probability measure μ_0 of the classical system.

An *interaction* of the quantum lattice system will be a function Φ from finite nonempty subsets X of Z^ν to self-adjoint observables $\Phi(X) \in \mathfrak{A}_X$, satisfying the translation-invariance condition

$$\Phi(X+i) = \tau_i \Phi(X) \quad \text{for all } i \in Z^\nu, X \subset Z^\nu \text{ finite nonempty}.$$

The Hamiltonian for the finite region Λ with interaction Φ is

$$H_\Lambda^\Phi = \sum_{X \subset \Lambda} \Phi(X)\,.$$

As in the classical case, we have the linear space $\mathcal{B}_0$ of finite-range interactions, and the Banach spaces $\mathcal{B}$ and $\tilde{\mathcal{B}}$ with their respective norms

$$|||\Phi||| = \sum_{X \ni 0} \frac{\|\Phi(X)\|}{|X|}$$

and

$$\|\Phi\|_{\sim} = \sum_{X \ni 0} \|\Phi(X)\| .$$

The classical lattice system with Ω_0 discrete and μ_0 normalized counting measure can be considered as a special case of the quantum system. We pick an orthonormal basis of $\mathcal{H}_0$ indexed by the points of Ω_0; for each finite $\Lambda \subset Z^\nu$, the $|\Lambda|$-fold tensor products of these basis elements form an orthonormal basis of $\mathcal{H}_\Lambda$. We can inject $C(\Omega_\Lambda)$ into $\mathfrak{A}_\Lambda$ as diagonal matrices with respect to this basis, and this extends to an injection of $C(\Omega)$ into $\mathfrak{A}$. An interaction of the classical system then corresponds to an interaction of the quantum system with each $\Phi(X)$ diagonal. Most of our results will have very similar formulations and proofs for classical and quantum systems.

The pressure for the finite region Λ in the quantum lattice system is given by

$$P_\Lambda(\Phi) = |\Lambda|^{-1} \ln \operatorname{tr} e^{-H_\Lambda^\Phi} . \tag{21}$$

We will use very similar techniques to deal with the infinite-volume limits of the pressure in classical and quantum systems. The quantum equivalent of Lemma I.2.1 is the following:

LEMMA 8.3.1. *Let* ρ *be any state on a* C^**-algebra* $\mathfrak{A}$ *with identity. Then for any self-adjoint* $B \in \mathfrak{A}$, $e^{\rho(B)} \leq \rho(e^B)$.

Proof. This is really just a statement about the abelian C^*-algebra generated by B; but that is isomorphic to the algebra of continuous functions on the spectrum of B, with the state ρ corresponding to a probability measure on the spectrum of B. This reduces the lemma to the classical case, Lemma I.2.1. ∎

LEMMA I.3.2. *For any self-adjoint* $n\times n$ *matrices* A, B:

$$|\ln \operatorname{tr} e^A| \le \|A\| \tag{22}$$

$$|\ln \operatorname{tr} e^A - \ln \operatorname{tr} e^B| \le \|A-B\| . \tag{23}$$

Proof. (22) is reduced to the classical case (13) in the same way as Lemma I.3.1 was reduced above to Lemma I.2.1.

Let $\{u_1, \cdots, u_n\}$ be an orthonormal basis of eigenvectors for A, with $Au_j = \lambda_j u_j$. Then $C \mapsto \langle u_j, Cu_j \rangle$ is a state on the C^*-algebra of complex $n\times n$ matrices for each j. Thus we have

$$\operatorname{tr} e^B = n^{-1} \sum_{j=1}^{n} \langle u_j, e^B u_j \rangle \ge n^{-1} \sum_{j=1}^{n} e^{\langle u_j, Bu_j \rangle}$$

$$\ge n^{-1} e^{-\|B-A\|} \sum_{j=1}^{n} e^{\lambda_j} = e^{-\|B-A\|} \operatorname{tr} e^A .$$

Taking logarithms, we have $\ln \operatorname{tr} e^B - \ln \operatorname{tr} e^A \ge -\|B-A\|$. To complete the proof we interchange A and B. ■

In the classical system, convexity of P was obtained using Hölder's inequality: $\langle e^{tf+(1-t)g} \rangle_0 \le \langle e^f \rangle_0^t \langle e^g \rangle_0^{1-t}$ for $0 \le t \le 1$, f and g real. Here is the version we use for quantum systems:

LEMMA I.3.3. *For self-adjoint* $n\times n$ *matrices* A, B *and* $0 \le t \le 1$,

$$\operatorname{tr} e^{tA+(1-t)B} \le (\operatorname{tr} e^A)^t (\operatorname{tr} e^B)^{1-t} . \tag{24}$$

Thus the function $A \mapsto \ln \operatorname{tr} e^A$ *on self-adjoint matrices is convex.*

Proof. Let $\{v_1, \cdots, v_n\}$ be an orthonormal basis of eigenvectors for $tA + (1-t)B$. Then

$$\operatorname{tr} e^{tA+(1-t)B} = n^{-1}\sum_{j=1}^{n} e^{t<v_j,Av_j>} e^{(1-t)<v_j,Bv_j>}$$

(by Holder's Inequality) $\leq \left(n^{-1}\sum_{j=1}^{n} e^{<v_j,Av_j>}\right)^{t}\left(n^{-1}\sum_{j=1}^{n} e^{<v_j,Bv_j>}\right)^{1-t}$

(by Lemma I.3.1) $\leq (\operatorname{tr} e^{A})^{t}(\operatorname{tr} e^{B})^{1-t}$. ∎

The same techniques used in the classical case, combined with the above inequalities, are used to prove the quantum versions of Theorems I.2.3 and I.2.4:

THEOREM I.3.4. *In the quantum lattice system, for* $\Phi \in \mathcal{B}$, $P(\Phi) = \lim_{a\to\infty} P_{C_a}(\Phi)$ *exists.* P *is a convex function on* $\mathcal{B}$ *with* $|P(\Phi)-P(\Psi)| \leq |||\Phi-\Psi|||$.

THEOREM I.3.5. *In the quantum lattice system, if* $\Lambda_n \to \infty$ *(van Hove) and* $\Phi \in \mathcal{B}$, *then* $P_{\Lambda_n}(\Phi) \to P(\Phi)$.

We now introduce boundary conditions for the quantum system. Recall that for the classical system, an external configuration was defined as a configuration τ in the complement of the finite region Λ in question. For an interaction of finite range, only that part of the configuration within a certain distance of Λ would make any contribution. The quantum analogue of this would be a vector v of unit norm in $\mathcal{H}_{\Lambda'}$ where $\Lambda' \subset \Lambda^c$ is finite, and for the interaction Φ in question, $\Phi(X) = 0$ when $X \cap \Lambda \neq \emptyset$ and $X \not\subset \Lambda \cup \Lambda'$. The Hamiltonian for the interaction Φ in Λ with the external configuration v would then be the operator ${}_vH^{\Phi}_{\Lambda}$ defined by

$$<w, {}_vH^{\Phi}_{\Lambda} u> = \sum_{X\cap\Lambda\neq\emptyset} <w\otimes v, \Phi(X)\, u\otimes v> \quad \text{for } w, u \in \mathcal{H}_{\Lambda}\,.$$

However, to deal with infinite-range interactions we must abandon the

Hilbert space and work with the C^*-algebras themselves. The configuration r and the vector v represent pure states (extreme points of the convex set of states) of the C^*-algebras $C(\Omega_{\Lambda^c})$ and $\mathfrak{A}_{\Lambda'}$ respectively. Thus we will define an *external configuration* in the quantum system for the finite subset Λ of Z^ν as a pure state h on $\mathfrak{A}_{\Lambda^c}$. Note that if id_Λ denotes the identity map on $\mathfrak{A}_\Lambda$, $id_\Lambda \otimes h$ is a contraction from $\mathfrak{A}$ to $\mathfrak{A}_\Lambda$. The Hamiltonian corresponding to the external configuration h is

$$(25) \qquad {}_hH^\Phi_\Lambda = \sum_{X \cap \Lambda \neq \emptyset} (id_\Lambda \otimes h)(\Phi(X)) .$$

For periodic boundary conditions, we again let $T_n = (Z/nZ)^\nu$, with the quotient map $q_n: Z^\nu \to T_n$, and the C^*-algebra $\mathfrak{A}_{T_n}$ of linear operators on $\mathcal{H}_{T_n} = \bigotimes_{i \in T_n} \mathcal{H}_i$, each $\mathcal{H}_i$ being a copy of $\mathcal{H}_0$. We have no analogue of the map $r_n: \Omega_{T_n} \to \Omega$ of the classical case; however, if q_n is one-to-one on $X \subset Z^\nu$ there is an induced isomorphism

$$i_X: \mathfrak{A}_X \to \mathfrak{A}_{q_n(X)} .$$

This is the reason for the restriction to such X in the definition of $\sum^*_{[X]}$. The periodic-boundary-condition Hamiltonian is thus

$$(26) \qquad H^\Phi_{T_n} = \sum^*_{[X]} i_X(\Phi(X)) .$$

As in the classical case, the sums in (25) and (26) converge for $\Phi \in \tilde{\mathcal{B}}$, with the estimates

$$\|{}_hH^\Phi_\Lambda\| \le |\Lambda| \, \|\Phi\|_\sim$$

$$\|H^\Phi_{T_n}\| \le n^\nu \|\Phi\|_\sim .$$

Just as in the classical case, we can now prove

THEOREM I.3.6. *In the quantum lattice system, let* $\Phi \in \tilde{\mathcal{B}}$. *Then the pressures* $P_{T_n}(\Phi)$ *for the periodic-boundary-condition Hamiltonians* $H^{\Phi}_{T_n}$ *converge to* $P(\Phi)$ *as* $n \to \infty$, *and for any sequence* $\Lambda_n \to \infty$ *(van Hove) with external configurations* h_n, *the pressures* ${}_{h_n}P_{\Lambda_n}(\Phi)$ *for Hamiltonians* ${}_{h_n}H^{\Phi}_{\Lambda_n}$ *also converge to* $P(\Phi)$.

The reader should be warned that the pressure as we have defined it differs from the more usual definitions for classical (discrete-spin) and quantum systems by $\ln|\Omega_0|$ and $\ln \dim \mathcal{H}_0$ respectively. This is because we have used the probability measure μ_0 and the normalized trace tr instead of summations over configurations and the usual trace Tr.

I.4. *Physical equivalence of interactions*

We will begin by comparing the spin and lattice-gas formulations for the classical system with $\Omega_0 = \{-1,+1\}$. The superscripts S and G will be used on our spaces $\mathcal{B}_0$, $\mathcal{B}$ and $\tilde{\mathcal{B}}$ to denote spaces of spin interactions (each $\Phi(X)$ a multiple of σ_X) and lattice-gas interactions (each $\Phi(X)$ a multiple of ρ_X) respectively. To avoid questions of convergence we will work with finite-range interactions.

Note that

$$\sigma_X = \prod_{i \in X} (2\rho_i - 1) = \sum_{Y \subset X} (-1)^{|X \setminus Y|} 2^{|Y|} \rho_Y$$

(27)

$$\rho_X = \prod_{i \in X} \tfrac{1}{2}(\sigma_i + 1) = 2^{-|X|} \sum_{Y \subset X} \sigma_Y .$$

A one-to-one correspondence between $\mathcal{B}_0^S$ and $\mathcal{B}_0^G$ can be set up as follows: suppose $\Phi^G \in \mathcal{B}_0^G$, so that $\Phi^G(X) = \tilde{\phi}(X)\rho_X$ for each X. Then $S\Phi^G \in \mathcal{B}_0^S$ is defined by $S\Phi^G(X) = \phi(X)\sigma_X$ with

(28) $$\phi(X) = \sum_{Y \supset X} 2^{-|Y|} \tilde{\phi}(Y) \quad (X \neq \emptyset) .$$

This can be inverted by the "inclusion-exclusion principle" to obtain

$$(29) \qquad \tilde{\phi}(X) = 2^{|X|} \sum_{Y \supset X} (-1)^{|Y \setminus X|} \phi(Y) \quad (X \neq \emptyset) .$$

The Hamiltonians for a finite subset Λ of Z^{ν} are

$$(30) \qquad H_{\Lambda}^{\Phi^G} = \sum_{X \subset \Lambda} \tilde{\phi}(X) \rho_X = \sum_{X \subset \Lambda} \sum_{Y \subset X} 2^{-|X|} \tilde{\phi}(X) \sigma_Y$$

$$(31) \qquad H_{\Lambda}^{S\Phi^G} = \sum_{Y \subset \Lambda} \phi(Y) \sigma_Y = \sum_{\substack{Y \subset \Lambda \\ Y \neq \emptyset}} \sum_{X \supset Y} 2^{-|X|} \tilde{\phi}(X) \sigma_Y .$$

The difference $H_{\Lambda}^{S\Phi^G} - H_{\Lambda}^{\Phi^G}$ consists of a constant term

$$(32) \qquad C_{\Lambda} = - \sum_{X \subset \Lambda} 2^{-|X|} \tilde{\phi}(X)$$

corresponding to $Y = \emptyset$ in (30), and a "boundary term"

$$(33) \qquad D_{\Lambda} = \sum_{\substack{Y \subset \Lambda \\ Y \neq \emptyset}} \sum_{\substack{X \supset Y \\ X \not\subset \Lambda}} 2^{-|X|} \tilde{\phi}(X) \sigma_Y .$$

Since Φ^G has finite range, the only nonzero contributions to (33) occur when Y is within a certain distance of the boundary of Λ.

For nonempty finite subsets X and Y of Z^{ν}, let $M_X^-(Y)$ be the number of translates of X contained in Y, and let $M_X^+(Y)$ be the number of translates of X which intersect Y. Then as $Y \to \infty$ (*van Hove*), for each X we will have $|Y|^{-1} M_X^-(Y) \to 1$ and $|Y|^{-1} M_X^+(Y) \to 1$.

We have the estimate

$$\|D_\Lambda\|_\infty \leq \sum_{\substack{X \cap \Lambda \neq \emptyset \\ X \not\subset \Lambda}} |\tilde{\phi}(X)|\, 2^{-|X|} \left\| \sum_{\substack{Y \subset X \cap \Lambda \\ Y \neq \emptyset}} \sigma_Y \right\|_\infty \leq \sum_{\substack{X \cap \Lambda \neq \emptyset \\ X \not\subset \Lambda}} |\tilde{\phi}(X)|$$

$$= \sum_{Z \ni 0} |Z|^{-1} (M_Z^+(\Lambda) - M_Z^-(\Lambda))\, |\tilde{\phi}(Z)|$$

so that $|\Lambda|^{-1} \|D_\Lambda\|_\infty \to 0$ as $\Lambda \to \infty$ (van Hove). We can express C_Λ by

$$C_\Lambda = - \sum_{Z \ni 0} |Z|^{-1} M_Z^-(\Lambda)\, 2^{-|Z|}\, \tilde{\phi}(Z)$$

so that as $\Lambda \to \infty$ (van Hove)

$$|\Lambda|^{-1} C_\Lambda \to C \equiv - \sum_{Z \ni 0} |Z|^{-1}\, 2^{-|Z|}\, \tilde{\phi}(Z) .$$

Therefore, for any $\Psi \in \mathcal{B}$ we obtain (using the estimate (17))

$$P(\Phi^G + \Psi) = P(S\Phi^G + \Psi) + C$$

and

$$P(\Psi + t(\Phi^G - S\Phi^G)) = P(\Psi) + tC \quad \text{for all real } t .$$

Thus the pressure is linear in the direction $\Phi^G - S\Phi^G$. We will later define invariant equilibrium states in terms of tangent functionals to the pressure: a linear functional α on $\mathcal{B}$ is said to be *tangent to* P *at* $\Phi \in \mathcal{B}$ if $P(\Phi+\Psi) \geq P(\Phi) + \alpha(\Psi)$ for all $\Psi \in \mathcal{B}$. Thus the interactions Φ^G and $S\Phi^G$ *will have the same tangent functionals, and the same invariant equilibrium states.* When we define equilibrium states in terms of the DLR equations (in Chapter III) we will find that Φ^G and $S\Phi^G$ have the same equilibrium states. Since, apart from the unimportant constant term C, the behavior of Φ^G and $S\Phi^G$ in the infinite-volume limit is identical, we can regard them as different ways of expressing the same basic interaction.

We now investigate the norms of Φ^G and $S\Phi^G$. We have

$$|||S\Phi^G||| \leq \sum_{X \ni 0} |X|^{-1} \sum_{Y \supset X} 2^{-|Y|} |\tilde{\phi}(Y)| = \sum_{Y \ni 0} 2^{-|Y|} |\tilde{\phi}(Y)| \sum_{\substack{X \subset Y \\ 0 \in X}} |X|^{-1}$$

$$\leq |||\Phi^G|||$$

since

$$\sum_{\substack{X \subset Y \\ 0 \in X}} |X|^{-1} = \sum_{n=1}^{|Y|} n^{-1} \binom{|Y|-1}{n-1} = |Y|^{-1} (2^{|Y|} - 1) .$$

Similarly

$$\|S\Phi^G\|_{\sim} \leq \sum_{X \ni 0} \sum_{Y \supset X} 2^{-|Y|} |\tilde{\phi}(Y)| \leq \tfrac{1}{2} \|\Phi^G\|_{\sim} .$$

However, the best estimate for the inverse map $S\Phi^G \mapsto \Phi^G$ is

$$|||\Phi^G||| \leq \sum_{X \ni 0} \sum_{Y \supset X} |X|^{-1} 2^{|X|} |\phi(Y)| = \sum_{Y \ni 0} (3^{|Y|} - 1) |Y|^{-1} |\phi(Y)|$$

where

$$(34) \quad \sum_{\substack{X \subset Y \\ 0 \ni X}} |X|^{-1} 2^{|X|} = \sum_{n=1}^{|Y|} n^{-1} 2^n \binom{|Y|-1}{n-1} = |Y|^{-1} \sum_{n=1}^{|Y|} 2^n \binom{|Y|}{n}$$

$$= |Y|^{-1} (3^{|Y|} - 1) .$$

Equality is attained in this estimate if e.g., $\phi(Y) \geq 0$ when $|Y|$ is even and $\phi(Y) \leq 0$ when $|Y|$ is odd. Thus although S is bounded from $\mathscr{B}^G$ to $\mathscr{B}^S$ and from $\tilde{\mathscr{B}}^G$ to $\tilde{\mathscr{B}}^S$, S^{-1} is unbounded even from $\tilde{\mathscr{B}}^S$ to $\mathscr{B}^G$.

This means that the spin formulation allows a wider class of interactions than does the lattice gas; while finite-range spin interactions have corresponding lattice-gas interactions, there are some interactions in $\tilde{\mathcal{B}}^S$ which have no lattice-gas counterpart in $\mathcal{B}^G$.

We can extend S to an operator which projects $\mathcal{B}_0$ on $\mathcal{B}_0^S$, where Φ and $S\Phi$ can again be considered as different versions of the same interaction. We will define S on all of $\tilde{\mathcal{B}}$, but it will not be bounded, even from $\tilde{\mathcal{B}}$ to $\mathcal{B}^S$. Under pointwise multiplication, Ω is a compact abelian group, with characters σ_Y, $Y \subset Z^\nu$ finite. Each $\Phi(X)$ can then be expanded in a Fourier series

$$\Phi(X) = \sum_{Y \subset X} a_Y^X \sigma_Y \quad \text{where} \quad a_Y^X = \langle \Phi(X) \sigma_Y \rangle_0 .$$

The translation invariance of Φ implies $a_Y^X = a_{Y+i}^{X+i}$ for $i \in Z^\nu$. Now for $\Phi \in \tilde{\mathcal{B}}$ we define

(35) $$S\Phi(Y) = \sum_{X \supset Y} a_Y^X \sigma_Y \quad (Y \neq \emptyset) .$$

This series converges absolutely, with $|S\Phi(Y)| \leq \sum_{X \supset Y} |\Phi(X)| \leq \|\Phi\|_{\sim}$ and $|S\Phi(Y)| \to 0$ as $Y \to \infty$. $S\Phi$ is a "spin language" interaction, although in general it is not in the Banach space $\mathcal{B}^S$. Of course, if $\Phi \in \mathcal{B}_0$, then $S\Phi \in \mathcal{B}_0^S$. The Hamiltonians for the finite subset Λ of Z^ν are

(36) $$H_\Lambda^\Phi = \sum_{X \subset \Lambda} \sum_{Y \subset X} a_Y^X \sigma_Y$$

(37) $$H_\Lambda^{S\Phi} = \sum_{\substack{Y \subset \Lambda \\ Y \neq \emptyset}} \sum_{X \supset Y} a_Y^X \sigma_Y .$$

This construction can be further generalized to deal with arbitrary

Ω_0 and μ_0. The notion of a spin interaction (each $\Phi(X)$ a multiple of σ_X) is replaced in the general case by the conditions

$$(38) \qquad \int_{\Omega_{Y^c}} \Phi(X)(\omega \times \omega')\mu_0(d\omega') = 0 \text{ for all } Y \subsetneq X, \omega \in \Omega_Y .$$

The superscript S will be used to denote spaces of interactions satisfying (38) for all X (it is easily seen that in the case of the Ising model this agrees with the previous definition of spin interactions). Now consider the operators $E_\Lambda: C(\Omega) \to C(\Omega_\Lambda)$ defined by

$$(39) \qquad E_\Lambda f(\omega) = \int_{\Omega_{\Lambda^c}} f(\omega \times \omega')\mu_0(d\omega') \text{ for } \omega \in \Omega_\Lambda$$

(here $E_\emptyset = \mu_0: C(\Omega) \to \mathbf{C}$). To a probabilist, $E_\Lambda f$ is the conditional expectation $E(f \mid \mathcal{S}_\Lambda)$ for f with respect to the Borel σ-algebra $\mathcal{S}_\Lambda$ of Ω_Λ, under the probability measure μ_0. A functional analyst would call E_Λ the orthogonal projection of $L^2(\Omega, \mu_0)$ on $L^2(\Omega_\Lambda, \mu_0)$. Note $E_\Lambda E_{\Lambda'} = E_{\Lambda \cap \Lambda'}$. The conditions (38) say $E_Y \Phi(X) = 0$ for $Y \subsetneq X$.

We now define operators

$$Q_X = \sum_{Y \subset X} (-1)^{|X \setminus Y|} E_Y \text{ for } X \subset \mathbf{Z}^\nu \text{ finite}$$

so that by the inclusion-exclusion principle $E_Y = \sum_{X \subset Y} Q_X$. It is easily seen that Q_X is the orthogonal projection of $L^2(\Omega, \mu_0)$ on the subspace of $L^2(\Omega_X, \mu_0)$ orthogonal to $L^2(\Omega_Y, \mu_0)$ for all $Y \subsetneq X$. As an operator on $C(\Omega)$ we have $\|Q_X\| \leq 2^{|X|}$. In the Ising model, Q_X is the orthogonal projection on σ_X and has norm one. Finally, S is defined on $\bar{\mathcal{B}}$ by

$$(40) \qquad S\Phi(Y) = \sum_{X \supset Y} Q_Y(\Phi(X)) \quad (Y \neq \emptyset) .$$

Then $S\Phi$ is an interaction satisfying (38) (although generally not in $\mathcal{B}^S$) with $\|S\Phi(Y)\|_\infty \leq 2^{|Y|}\|\Phi\|_\sim$. The Hamiltonians for a finite subset Λ of Z^ν are

$$H_\Lambda^\Phi = \sum_{X\subset\Lambda} \sum_{Y\subset X} Q_Y(\Phi(X)) \tag{41}$$

$$H_\Lambda^{S\Phi} = \sum_{\substack{Y\subset\Lambda \\ Y\neq\emptyset}} \sum_{X\supset Y} Q_Y(\Phi(X)) . \tag{42}$$

The difference $H_\Lambda^{S\Phi} - H_\Lambda^\Phi$ consists again of constant term C_Λ and boundary term D_Λ, with

$$C_\Lambda = - \sum_{X\subset\Lambda} \langle\Phi(X)\rangle_0$$

$$D_\Lambda = \sum_{\substack{X\cap\Lambda\neq\emptyset \\ X\not\subset\Lambda}} \sum_{\substack{Y\subset X\cap\Lambda \\ Y\neq\emptyset}} Q_Y(\Phi(X)) = \sum_{\substack{X\cap\Lambda\neq\emptyset \\ X\not\subset\Lambda}} (E_{X\cap\Lambda}\Phi(X) - \langle\Phi(X)\rangle_0) .$$

Thus

$$\|D_\Lambda\|_\infty \leq 2 \sum_{\substack{X\cap\Lambda\neq\emptyset \\ X\not\subset\Lambda}} \|\Phi(X)\|_\infty \leq 2 \sum_{X\ni 0} |X|^{-1} (M_X^+(\Lambda) - M_X^-(\Lambda)) \|\Phi(X)\|_\infty$$

so that $|\Lambda|^{-1}\|D_\Lambda\|_\infty \to 0$ as $\Lambda \to \infty$ (van Hove), while

$$|\Lambda|^{-1} C_\Lambda = - \sum_{X\ni 0} |X|^{-1} |\Lambda|^{-1} M_X^-(\Lambda)\langle\Phi(X)\rangle_0 \to -\langle A_\Phi\rangle_0$$

where $A_\Phi = \sum_{X\ni 0} \frac{\Phi(X)}{|X|}$ represents the contribution of the site 0 to the energy. The expectation of A_Φ in any state is the average energy per site in that state. Now if $S\Phi \in \mathcal{B}^S$ and $\Psi \in \mathcal{B}$ we have

$$P(S\Phi+\Psi) = P(\Phi+\Psi) + \langle A_\Phi \rangle_0$$

so that Φ and $S\Phi$ have the same invariant equilibrium states. We will say that two interactions $\Phi, \Psi \in \mathcal{B}$ are *physically equivalent* if $\Phi - \Psi \in \tilde{\mathcal{B}}$ and $S(\Phi-\Psi) = 0$. Note that Φ and Ψ themselves need not be in $\tilde{\mathcal{B}}$, so $S\Phi$ and $S\Psi$ may be undefined. We then have

THEOREM I.4.1. *In the classical lattice system, if* Φ *and* Ψ *are physically equivalent interactions in* $\mathcal{B}$, *then* $P(\Phi) + \langle A_\Phi \rangle_0 = P(\Psi) + \langle A_\Psi \rangle_0$, *and* Φ *and* Ψ *have the same invariant equilibrium states.*

An estimate for $|||S\Phi|||$ can be obtained as follows:

$$|||S\Phi||| \le \sum_{Y \ni 0} \sum_{X \supset Y} |Y|^{-1} \|Q_Y(\Phi(X))\|_\infty$$

$$\le \sum_{X \ni 0} \|\Phi(X)\|_\infty \sum_{\substack{Y \subset X \\ 0 \in Y}} |Y|^{-1} 2^{|Y|}$$

$$= \sum_{X \ni 0} |X|^{-1} (3^{|X|} - 1) \|\Phi(X)\|_\infty$$

using (34). For the Ising model a somewhat better estimate can be obtained. It is still not obvious that S is unbounded from $\tilde{\mathcal{B}}$ to $\mathcal{B}^S$. To show this, let $\Lambda \subset Z^\nu$ with $|\Lambda| = N$, containing at most one translate of any set of more than one lattice point (e.g., with $\nu = 1$, take $\Lambda = \{3^n: n = 1,2,\cdots,N\}$). We will take the interaction Ψ_A^Λ defined in (9) of Section I.1 where $A \in C(\Omega_\Lambda)$ is real-valued, $\|A\|_\infty = 1$. Thus $\|\Psi_A^\Lambda\|_{\sim} = N$. For simplicity we work in the Ising model, so that $A = \sum_{X \subset \Lambda} \hat{A}(X)\sigma_X$. Then

$$S\Psi_A^\Lambda(\{i\}) = \sum_{j \in \Lambda} \hat{A}(\{j\})\sigma_i$$

and for $|X| > 1$

$$S\Psi_A^{\Lambda}(X) = \begin{cases} \hat{A}(i+X)\sigma_X & \text{if } i+X \subset \Lambda \\ 0 & \text{if } \Lambda \text{ contains no translate of } X . \end{cases}$$

Thus

$$|||S\Psi_A^{\Lambda}||| = \left|\sum_{j\in\Lambda} \hat{A}(\{j\})\right| + \sum_{\substack{X\subset\Lambda \\ |X|>1}} |\hat{A}(X)| .$$

Now instead of providing a specific A we use a probabilistic argument. Let $\{A(\omega): \omega \in \Omega_\Lambda\}$ be independent random variables, each taking values ± 1 with probability $\frac{1}{2}$. Each $\hat{A}(X) = \langle A\, \sigma_X \rangle_0$ is then an average of 2^N independent random variables $A(\omega)\,\sigma_X(\omega)$, $\omega \in \Omega_\Lambda$. By the standard methods of probability theory

$$2^{N/2}\, E(|\hat{A}(X)|) \to \sqrt{\tfrac{2}{\pi}} \quad \text{as } N \to \infty$$

where E stands for expectation (of course $E(|\hat{A}(X)|)$ is the same for each X). So for N large, $E(|\hat{A}(X)|) \geq 2^{-(N/2)-1}$ and

$$E\left(\sum_{\substack{X\subset\Lambda \\ |X|>1}} |\hat{A}(X)|\right) = (2^N - N - 1)\, E(|\hat{A}(X)|) > 2^{(N/2)-1} - 1 .$$

Therefore, for at least some A, $|||S\Psi_A^{\Lambda}||| > 2^{(N/2)-1} - 1$, and S is not bounded from $\tilde{\mathfrak{B}}$ to $\mathfrak{B}^S$.

In the quantum lattice system, the analogue of the operator E_Λ is the normalized partial trace tr_{Λ^c} (i.e., $\mathrm{id}_\Lambda \otimes \mathrm{tr}_{\Lambda^c}: \mathfrak{A} \to \mathfrak{A}_\Lambda$ where id_Λ is the identity map on $\mathfrak{A}_\Lambda$ and tr_{Λ^c} is the normalized trace on $\mathfrak{A}_{\Lambda^c}$). The superscript S will be used to denote spaces of quantum interactions with $\mathrm{tr}_Y(\Phi(X)) = 0$ for all $Y \subset X$, $Y \neq \emptyset$. By analogy with the classical system, we have the operators

$$Q_X = \sum_{Y \subset X} (-1)^{|X \setminus Y|} \operatorname{tr}_{Y^c} : \mathfrak{A} \to \mathfrak{A}_X$$

and we can define the operator S on $\mathcal{B}$ as in (40). The same notion of physical equivalence applies, and the same proof as in the classical case shows

THEOREM I.4.2. *In the quantum lattice system, if* Φ *and* Ψ *are physically equivalent interactions in* $\mathcal{B}$, *then* $P(\Phi) + \operatorname{tr} A_\Phi = P(\Psi) + \operatorname{tr} A_\Psi$, *and* Φ *and* Ψ *have the same invariant equilibrium states.*

It should be noted that our definition of physical equivalence of interactions differs somewhat from that of Roos [23]; his definition applies to quantum interactions in a more restrictive space of interactions (on which time evolution is defined), and requires the additional condition $\operatorname{tr} A_\Phi = \operatorname{tr} A_\Psi$. We will continue our study of physical equivalence in Chapter III, using the DLR equations and KMS conditions.

II. TANGENT FUNCTIONALS AND THE VARIATIONAL PRINCIPLE

II.1. P-*bounded functionals*

In this chapter we will define invariant equilibrium states of our classical and quantum lattice systems in terms of tangent functionals to the pressure on the space $\mathfrak{B}$ of interactions, and show that this is equivalent to an alternative definition by the Variational Principle: invariant equilibrium states for a given interaction maximize the difference between entropy and energy per site (recall that we are absorbing the temperature into the interaction, so that $\beta = 1$). The first step is to develop the connection between translation-invariant states and functionals on the space of interactions.

A *state* on a C^*-algebra $\mathfrak{A}$ with identity is a positive linear functional ρ on $\mathfrak{A}$ (i.e., $\rho(A) \geq 0$ if $A \geq 0$) with the normalization $\rho(1) = 1$. A state of the quantum lattice system is a state on the quasi-local C^*-algebra $\mathfrak{A}$; a state of the classical lattice system is a state on $C(\Omega)$, which by the Riesz Representation Theorem is identified with a probability measure on the configuration space Ω. We will mainly consider translation-invariant states, i.e., states ρ with $\rho \circ \tau_i = \rho$ for all $i \in Z^\nu$. The set of translation-invariant states will be denoted E^I; it is a convex set, compact in the weak-* topology.

If P is any convex function on a Banach space $\mathfrak{X}$, a linear functional $\alpha \in \mathfrak{X}^*$ is said to be *tangent to* P *at* $\Phi \in \mathfrak{X}$ if for all $\Psi \in \mathfrak{X}$

$$P(\Phi + \Psi) \geq P(\Phi) + \alpha(\Psi) . \tag{1}$$

We will say that the functional α is P-*bounded* if for some constant

C, $P(\Psi) \geq \alpha(\Psi) + C$ for all $\Psi \in \mathfrak{X}$. Thus tangent functionals are P-bounded functionals α such that $P-\alpha$ attains its infimum at some $\Phi \in \mathfrak{X}$.

We consider the map from invariant states ρ to linear functionals $\alpha \in \mathcal{B}^*$ defined by

$$(2) \qquad \alpha(\Psi) = -\rho(A_\Psi) \text{ for } \Psi \in \mathcal{B}$$

where $A_\Psi = \sum_{X \ni 0} |X|^{-1}\Psi(X)$. An *invariant equilibrium state* for the interaction $\Phi \in \mathcal{B}$ is an invariant state ρ such that the functional α defined by (2) is tangent to the pressure P at Φ.

LEMMA II.1.1. *The map from* E^I *to* $\mathcal{B}^*$ *defined by* (2) *is a homeomorphism in the weak-* topologies, and an isometry for the respective norms.*

Proof. For simplicity we will treat the classical case. It is clear from (2) that the map is continuous in the weak-* topologies, and (since $\|A_\Psi\|_\infty \leq |||\Psi|||$) that $\|\alpha-\alpha'\| \leq \|\rho-\rho'\|$ where α and α' are the images of ρ and ρ' respectively under the map. Since E^I is compact in the weak-* topology, the map is a homeomorphism if it is one-to-one. We have to show that $\|\alpha-\alpha'\| \geq \|\rho-\rho'\|$. For $A \in C(\Omega_\Lambda)$ real-valued with Λ finite, we consider the interaction Ψ_A^Λ defined as in formula (9) of Section I.1. Then $|||\Psi_A^\Lambda||| = \|A\|_\infty$ and $\rho(A_{\Psi_A^\Lambda}) = \rho(A)$ for any $\rho \in E^I$, so that

$$|\rho(A)-\rho'(A)| = |\alpha(\Psi_A^\Lambda)-\alpha'(\Psi_A^\Lambda)| \leq \|\alpha-\alpha'\| \, \|A\|_\infty .$$

Since the union of the local algebras $C(\Omega_\Lambda)$ is dense in $C(\Omega)$, $\|\rho-\rho'\| \leq \|\alpha-\alpha'\|$. The proof of the quantum case is similar, with one technicality: we must show that the norm of a hermitian linear functional h on a C^*-algebra is equal to the norm of its restriction to self-adjoint elements (in the classical case this is trivial by the Riesz Representation

Theorem). If $|h(A)| \leq c\|A\|$ for A self-adjoint and if T is any element of the algebra, we can assume (multiplying by a complex number) that h(T) is real, and then

$$|h(T)| = |h(\text{Re } T)| \leq c\|\text{Re } T\| \leq c\|T\| . \qquad \blacksquare$$

If $\mathcal{B}$ is replaced by another Banach space of interactions with a norm dominating $|||\cdot|||$ (such as $\|\cdot\|_{\sim}$) and containing interactions physically equivalent to each interaction in $\mathcal{B}_0$, then the map from E^I to the dual of the new Banach space defined as in (2) will still be a homeomorphism in the weak-* topologies, but will no longer be an isometry. From this point of view, $\mathcal{B}$ is the most "natural" Banach space for considering invariant equilibrium states as defined by tangent functionals. However, in Chapter V we will see some very strange consequences of this "naturalness."

The next question of interest is to study the image of E^I under the map (2); in particular, we want to show that every tangent functional to P arises from an invariant state. It turns out that P-boundedness is enough to ensure that a functional is in the image of the map (2).

THEOREM II.1.2. *If $a \in \mathcal{B}^*$ is P-bounded, then there is a unique invariant state ρ with $a(\Psi) = -\rho(A_\Psi)$ for $\Psi \in \mathcal{B}$.*

Proof. Uniqueness is clear because the map defined by (2) is one-to-one. Again we treat the classical case for simplicity; the proof in the quantum case is similar.

Let $\Lambda \subset Z^\nu$ be finite, and for $A \in C(\Omega_\Lambda)$ consider the interaction Ψ_A^Λ. A linear functional ρ_Λ on $C(\Omega_\Lambda)$ is defined by $\rho_\Lambda(A) = -a(\Psi_A^\Lambda)$. We must show that ρ_Λ is a state on $C(\Omega_\Lambda)$ (i.e., a probability measure on Ω_Λ).

Consider first the constant function 1. Then for any finite Λ', $H_{\Lambda'}^{\Psi_1^\Lambda}$ is the constant function with value $M_\Lambda^-(\Lambda')$, the number of translates of

Λ contained in Λ'. Thus

$$P_{\Lambda'}(t\Psi_1^{\Lambda}) = |\Lambda'|^{-1} \ln e^{-tM_{\Lambda}^{-}(\Lambda')} = -t|\Lambda'|^{-1}M_{\Lambda}^{-}(\Lambda')$$

$$\to -t \text{ as } \Lambda' \to \infty \text{ (van Hove), for all real } t.$$

Since α is P-bounded, for some C we have $t\alpha(\Psi_1^{\Lambda}) + C \leq -t$ for all real t, and this implies $\alpha(\Psi_1^{\Lambda}) = -1$, so $\rho_{\Lambda}(1) = 1$.

Now suppose $A \geq 0$. Then $H^{\Psi_A^{\Lambda}} \geq 0$, and we obtain $P(t\Psi_A^{\Lambda}) \leq 0$ for $t \geq 0$. The P-boundedness condition $t\,\alpha(\Psi_A^{\Lambda}) + C \leq 0$ for all $t \geq 0$ implies $\alpha(\Psi_A^{\Lambda}) \leq 0$, so $\rho_{\Lambda}(A) \geq 0$.

Next we show that the ρ_{Λ} are consistent. Suppose $\Lambda' \supset \Lambda$ is finite, and $A \in C(\Omega_{\Lambda})$. Then $\Psi_A^{\Lambda'}$ and Ψ_A^{Λ} are physically equivalent: we have $S\Psi_A^{\Lambda}(X) = \sum_{i+\Lambda \supset X} Q_X(\tau_i A)$ and $S\Psi_A^{\Lambda'}(X) = \sum_{i+\Lambda' \supset X} Q_X(\tau_i A)$, but $Q_X(\tau_i A) = 0$ unless $X \subset i+\Lambda$. Moreover, $\langle A_{\Psi_A^{\Lambda}} \rangle_0 = \langle A_{\Psi_A^{\Lambda'}} \rangle_0 = \langle A \rangle_0$. By Theorem I.4.1, $P(t(\Psi_A^{\Lambda'} - \Psi_A^{\Lambda})) = 0$ (this is also easy to prove directly) for all real t. Thus the P-boundedness of α implies $\alpha(\Psi_A^{\Lambda'} - \Psi_A^{\Lambda}) = 0$, so $\rho_{\Lambda'}(A) = \rho_{\Lambda}(A)$.

Finally, the consistent family $\{\rho_{\Lambda}\}$ extends uniquely to a state ρ on $C(\Omega)$ (a probability measure on Ω). It is clear from the method of definition that ρ is invariant. Since for $A \in C(\Omega_{\Lambda})$, $A_{\Psi_A^{\Lambda}}$ is an average of translates of A, we obtain $\rho(A_{\Psi}) = -\alpha(\Psi)$, first for finite-range interactions and then by continuity for all $\Psi \in \mathcal{B}$. ■

The invariant states ρ which correspond to P-bounded functionals by (2) are those with $\inf\{P(\Phi)+\rho(A_{\Phi}) : \Phi \in \mathcal{B}\} > -\infty$. This quantity will turn out to be the mean entropy of ρ.

Although we stated Theorem II.1.2 for the space $\mathcal{B}$, it clearly remains true for any Banach space $\mathcal{B}_1$ dense in $\mathcal{B}$ with a norm dominating $|||\cdot|||$: the P-boundedness condition on a functional implies that it is in

$\mathcal{B}^*$ since $|P(\Psi)| \leq |||\Psi|||$ for all $\Psi \in \mathcal{B}$. Thus the P-bounded linear functionals on $\mathcal{B}$ are unique extensions of those on $\mathcal{B}_1$. Moreover, by Theorems I.4.1 and I.4.2 it is easily seen that for any P-bounded linear functional α on $\mathcal{B}$ and any $\Phi, \Psi \in \mathcal{B}$ physically equivalent, $\alpha(\Psi - \Phi) = \langle A_\Phi - A_\Psi \rangle_0$ (in the classical case) or $\operatorname{tr}(A_\Phi - A_\Psi)$ (in the quantum case). By this identity, any P-bounded functional on $\mathcal{B}^S$ has a unique P-bounded extension to $\mathcal{B}$. Thus the theorem is also true for $\mathcal{B}^S$ or any of its dense subspaces.

It is also of interest to consider states and interactions which are invariant under a larger symmetry group than Z^ν. The most important case occurs not in lattice systems but in hard-core continuous systems (which can be treated by similar methods), where one is concerned not only with translation invariance (in R^ν in this case) but with Euclidean invariance [9]. We will consider a compact group H which acts as follows:

(a) on Z^ν by linear transformations $i \mapsto h(i)$.

(b) (in the classical case) by μ_0-preserving automorphisms τ_h of $C(\Omega)$ taking $C(\Omega_\Lambda)$ to $C(\Omega_{h(\Lambda)})$, such that $h \mapsto \tau_h$ is strongly continuous and $\tau_h \circ \tau_j \circ \tau_h^{-1} = \tau_{h(j)}$ for $h \in H$ and $j \in Z^\nu$. These will be induced by continuous μ_0-preserving transformations T_h of Ω_0.

(c) (in the quantum case) by automorphisms and possibly anti-automorphisms τ_h of $\mathfrak{A}$ taking $\mathfrak{A}_\Lambda$ to $\mathfrak{A}_{h(\Lambda)}$, with $h \mapsto \tau_h$ strongly continuous and $\tau_h \circ \tau_j \circ \tau_h^{-1} = \tau_{h(j)}$. *An anti-automorphism* is a linear map τ_h with $\tau_h(AB) = \tau_h(B)\tau_h(A)$ for $A, B \in \mathfrak{A}$. It is easily seen that isomorphisms or anti-isomorphisms from $\mathfrak{A}_\Lambda$ to $\mathfrak{A}_{h(\Lambda)}$ must preserve the trace. The τ_h can be induced by unitary (for automorphisms) or anti-unitary (for anti-automorphisms) *transformations* U_h of $\mathcal{H}_0$.

(d) on states by $\rho \mapsto \rho \circ \tau_h^{-1}$, and on interactions by $\Psi^h(X) = \tau_h \Psi(h^{-1}(X))$.

We let E^H and $\mathcal{B}^H$ be the spaces of states and interactions (in $\mathcal{B}$) respectively, which are invariant under both Z^ν and H.

COROLLARY II.1.3. *If* $\alpha \in (\mathcal{B}^H)^*$ *is* P-*bounded, there is a unique* $\rho \in E^H$ *such that* $\alpha(\Psi) = -\rho(A_\Psi)$ *for* $\Psi \in \mathcal{B}^H$. *Moreover,*

$$\inf\{P(\Psi)+\rho(A_\Psi): \Psi \in \mathcal{B}\} = \inf\{P(\Psi)-\alpha(\Psi): \Psi \in \mathcal{B}^H\}.$$

Thus $\rho \in E^H$ *is an invariant equilibrium state for* $\Phi \in \mathcal{B}^H$ *if and only if* α *is tangent to* P *(restricted to* $\mathcal{B}^H$*) at* Φ.

Proof. Suppose $C = \inf\{P(\Psi)-\alpha(\Psi): \Psi \in \mathcal{B}^H\}$. By the Hahn-Banach theorem there is an extension α' of α to a linear functional on $\mathcal{B}$ with $P(\Psi) \geq \alpha'(\Psi) + C$ for all $\Psi \in \mathcal{B}$. Let $\alpha'' \in \mathcal{B}^*$ be defined by $\alpha''(\Psi) = \int_H \alpha'(\Psi^h)\,dh$ where dh is normalized Haar measure on H. Note that $P(\Psi^h) = P(\Psi)$, so $P(\Psi) \geq \alpha''(\Psi) + C$ for all $\Psi \in \mathcal{B}$. Then α'' determines a translation-invariant state ρ by Theorem II.1.2, and it is easily seen that ρ has the desired properties. To show uniqueness, suppose ρ and ρ' are distinct members of E^H. Then they must differ on some $A \in C(\Omega_\Lambda)$ (or $\mathfrak{A}_\Lambda$ in the quantum system) with Λ finite. Taking $\Psi' = \int_H (\Psi_A^\Lambda)^h\,dh \in \mathcal{B}^H$ we have $\rho(A_{\Psi'}) = \rho(A) \neq \rho'(A) = \rho'(A_{\Psi'})$. ∎

II.2. *The mean entropy*

The next quantity to study in the infinite-volume limit is the entropy per site. For a finite set $\Lambda \subset Z^\nu$ the entropy $S_\Lambda(\rho)$ will be a function on states ρ (or rather on their restrictions to $C(\Omega_\Lambda)$ or $\mathfrak{A}_\Lambda$) with values in $[-\infty, 0]$. The *mean entropy* $s(\rho)$ will be the limit of $|\Lambda|^{-1} S_\Lambda(\rho)$ as $\Lambda \to \infty$ (van Hove), and will be defined on the space E^I of invariant states (it will also exist for periodic states, i.e., states ρ with $\rho = \rho \circ \tau_j$ for j in some ν-dimensional subgroup of Z^ν). In keeping with our policy of using a probability measure on Ω_0 instead of counting measure in the classical system and the normalized trace tr instead of the usual traces Tr_Λ in the quantum system, our definitions of entropy will differ (by harmless constants) from the usual definitions.

Consider a state ρ of the classical lattice system, and a finite subset Λ of Z^{ν}. The restriction of ρ to $C(\Omega_\Lambda)$ defines a probability measure on Ω_Λ. If this measure is absolutely continuous with respect to μ_0, with Radon-Nikodym derivative $\rho^{(\Lambda)} \in L^1(\Omega_\Lambda, \mu_0)$ (so that for $A \in C(\Omega_\Lambda)$, $\rho(A) = <\rho^{(\Lambda)} A>_0$), we define the *entropy of* ρ *in* Λ by

$$S_\Lambda(\rho) = -\rho(\ln \rho^{(\Lambda)}) = -<\rho^{(\Lambda)} \ln \rho^{(\Lambda)}>_0 . \tag{3}$$

Since $t \ln t \geq t - 1$ for $t \geq 0$, $S_\Lambda(\rho)$ is well-defined and nonpositive (but possibly $-\infty$). If the restriction of ρ to $C(\Omega_\Lambda)$ is not absolutely continuous with respect to μ_0 we define $S_\Lambda(\rho) = -\infty$.

In the cases where Ω_0 is finite with normalized counting measure μ_0, any measure on Ω_Λ is absolutely continuous with respect to μ_0; moreover $\rho^{(\Lambda)} \leq |\Omega_0|^{|\Lambda|}$, so that $S_\Lambda(\rho) \geq -|\Lambda| \ln |\Omega_0|$. The more usual definition of entropy for these systems would work with $\rho_\Lambda = |\Omega_0|^{-|\Lambda|} \rho^{(\Lambda)}$ instead of $\rho^{(\Lambda)}$ (so that $\rho(A) = \sum_{\omega \in \Omega_\Lambda} \rho_\Lambda(\omega) A(\omega)$ for $A \in C(\Omega_\Lambda)$); the entropy of ρ in Λ would then be $S'_\Lambda(\rho) = -\sum_{\omega \in \Omega_\Lambda} \rho_\Lambda(\omega) \ln \rho_\Lambda(\omega)$, so that $S'_\Lambda(\rho) = S_\Lambda(\rho) + |\Lambda| \ln |\Omega_0|$.

The properties of our version of entropy resemble those of classical continuous systems (as in [21]). For example, our entropy is nonpositive, while the usual lattice gas and quantum versions are nonnegative (see [25], Section 7.2). We begin with some fundamental properties of $S_\Lambda(\rho)$. With ρ fixed, we will simplify the notation by using $S(\Lambda)$ to denote $S_\Lambda(\rho)$.

LEMMA II.2.1. *For any probability measure* ρ *on* Ω *and finite subsets* Λ, Λ' *of* Z^{ν} *we have*

(a) $S(\Lambda) \leq 0$ (*negativity*)

(b) $\Lambda \subseteq \Lambda' \Rightarrow S(\Lambda') \leq S(\Lambda)$ (*decrease*)

(c) $S(\Lambda \cup \Lambda') + S(\Lambda \cap \Lambda') \leq S(\Lambda) + S(\Lambda')$ (*strong subadditivity*).

Proof. Let $\Lambda \subseteq \Lambda'$. Then (assuming the Radon-Nikodym derivatives exist)

$$S(\Lambda') - S(\Lambda) = \rho(\ln \rho^{(\Lambda)} - \ln \rho^{(\Lambda')}) \leq \rho\left(\frac{\rho^{(\Lambda)}}{\rho^{(\Lambda')}} - 1\right)$$

where we have used the inequality $\ln t \leq t-1$ with $t = \rho^{(\Lambda)}/\rho^{(\Lambda')}$. The function $\rho^{(\Lambda)}/\rho^{(\Lambda')}$ need only be considered on the set $\{\omega : \rho^{(\Lambda')}(\omega) > 0\}$, which has full ρ-measure. Now $\rho(\rho^{(\Lambda)}/\rho^{(\Lambda')}) = <\rho^{(\Lambda)}>_0 = \rho(1) = 1$, so $S(\Lambda') - S(\Lambda) \leq 0$. Taking in particular $\Lambda = \emptyset$ (interpreting $\rho^{(\emptyset)} \equiv 1$) we also obtain (a). For (c),

$$S(\Lambda \cup \Lambda') - S(\Lambda) - S(\Lambda') + S(\Lambda \cap \Lambda') = \rho\left(\ln \frac{\rho^{(\Lambda)}\rho^{(\Lambda')}}{\rho^{(\Lambda \cup \Lambda')}\rho^{(\Lambda \cap \Lambda')}}\right)$$

where we only consider the subset of $\Omega_{\Lambda \cup \Lambda'}$ where $\rho^{(\Lambda \cup \Lambda')} > 0$. Again using $\ln t \leq t-1$, we obtain

$$S(\Lambda \cup \Lambda') - S(\Lambda) - S(\Lambda') + S(\Lambda \cap \Lambda') \leq \rho\left(\frac{\rho^{(\Lambda)}\rho^{(\Lambda')}}{\rho^{(\Lambda \cup \Lambda')}\rho^{(\Lambda \cap \Lambda')}}\right) - 1 =$$

$$= <\rho^{(\Lambda)}\rho^{(\Lambda')}/\rho^{(\Lambda \cap \Lambda')}>_0 - 1 = 0$$

since $<\rho^{(\Lambda')}(\cdot \times S)>_{0,\Lambda' \setminus \Lambda} = \rho^{(\Lambda' \cap \Lambda)}(S)$ for $S \in \Omega_{\Lambda' \cap \Lambda}$.

A weaker form of strong subadditivity is called "subadditivity": $S(\Lambda \cup \Lambda') \leq S(\Lambda) + S(\Lambda')$ for Λ and Λ' disjoint. This will be much easier to obtain in the quantum system than strong subadditivity.

THEOREM II.2.2. *In the classical lattice system, for* $\rho \in E^I$ *the mean entropy*

(4) $$s(\rho) = \lim_{\Lambda \to \infty \text{ (van Hove)}} |\Lambda|^{-1} S_\Lambda(\rho)$$

exists (in $[-\infty, 0]$*).*

Proof. Using subadditivity, it is easy to see that for a sequence of cubes C_{2^n} of side 2^n, the sequence $2^{-n\nu}S(C_{2^n})$ decreases as $n \to \infty$, and thus tends to a limit $s(\rho)$ (possibly $-\infty$). Now consider any sequence $\Lambda_n \to \infty$ (van Hove), and let a be a fixed power of 2. As in the remarks before Theorem I.2.4 we can take $\Lambda_n^- \subseteq \Lambda_n$, where Λ_n^- in the union of $N_a^-(\Lambda_n)$ disjoint cubes of side a, and $|\Lambda_n|^{-1} N_a^-(\Lambda_n) \to a^{-\nu}$ as $n \to \infty$. Then

$$\limsup_{n\to\infty} |\Lambda_n|^{-1} S(\Lambda_n) \leq \limsup_{n\to\infty} |\Lambda_n|^{-1} S(\Lambda_n^-)$$
$$\leq \lim_{n\to\infty} |\Lambda_n|^{-1} N_a^-(\Lambda_n) S(C_a) = a^{-\nu} S(C_a) .$$

As $a \to \infty$ we have $\limsup_{n\to\infty} |\Lambda_n|^{-1} S(\Lambda_n) \leq s(\rho)$. If $s(\rho) = -\infty$ this gives us (4). Assume now $s(\rho) > -\infty$.

This time, take $\Lambda_n \subseteq \Lambda_n^+$, where Λ_n^+ is the union of $N_a^+(\Lambda_n)$ disjoint cubes $X_j \in \mathcal{C}_a$ of side a. These will be taken in "lexicographic order." Let $Y_k = \bigcup_{j=1}^{k} X_j$. We have $S(X_1) \geq a^\nu s(\rho)$. We claim that

$$S(Y_{k+1}) - S(Y_k) \geq a^\nu s(\rho) \text{ for all } k .$$

If the claim is true, then $S(\Lambda_n) \geq S(\Lambda_n^+) \geq N_a^+(\Lambda_n) a^\nu s(\rho)$ so that $\liminf_{n\to\infty} |\Lambda_n|^{-1} S(\Lambda_n) \geq \lim_{n\to\infty} |\Lambda_n|^{-1} N_a^+(\Lambda_n) a^\nu s(\rho) = s(\rho)$ which will complete the proof.

Suppose the claim is false, so that $S(Y_{k+1}) - S(Y_k) \leq a^\nu (s(\rho) - \varepsilon)$ for some k and some $\varepsilon > 0$. Now by the strong subadditivity property, $S(Z \cup Y_{k+1}) - S(Z \cup Y_k) \leq S(Y_{k+1}) - S(Y_k) \leq a^\nu(s(\rho) - \varepsilon)$ whenever Z is disjoint from Y_{k+1}. Consider a large cube of side La (as in Figure 2) with a "border" of width $da \geq \operatorname{diam}(Y_{k+1})$ on all faces, such that Y_k is contained in the border region B with X_{k+1} in the "lower left corner" of the inner cube of side $(L-2d)a$. After adding X_{k+1} to B, with change in entropy $S(B \cup X_{k+1}) - S(B) \leq a^\nu (s(\rho) - \varepsilon)$, we have $Y_k + (a, 0, \cdots, 0) \subset B \cup X_{k+1}$. Then the cube $X' = X_{k+1} + (a, 0, \cdots, 0)$ can be

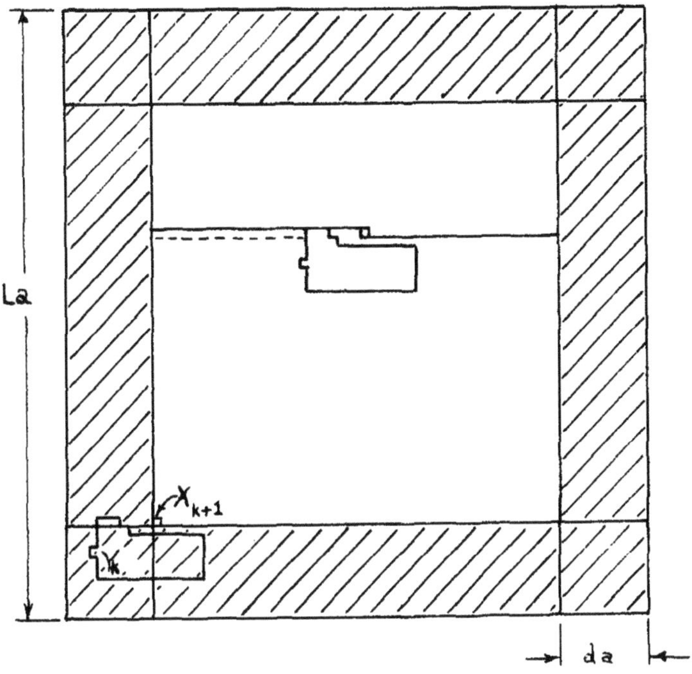

Figure 2

added on, with $S(B \cup X_{k+1} \cup X') - S(B \cup X_{k+1}) \leq a^{\nu}(s(\rho) - \varepsilon)$. Continuing the process, we can successively add cubes of side a in lexicographic order until the whole inner cube is filled, with an entropy difference of at most $a^{\nu}(s(\rho)-\varepsilon)$ each time. Thus we obtain $S(C_{La}) \leq S(B) + (L-2d)^{\nu} \times a^{\nu}(s(\rho)-\varepsilon) \leq (L^{\nu} - (L-2d)^{\nu}) S(C_a) + (L-2d)^{\nu} a^{\nu}(s(\rho)-\varepsilon)$. Dividing by $(La)^{\nu}$ and then taking the limit as $L \to \infty$ through powers of 2, this would yield the contradiction $s(\rho) < s(\rho) - \varepsilon$. This proves the claim. ■

THEOREM II.23. *The function* $\rho \mapsto s(\rho)$ *is affine and upper semicontinuous on* E^I *(with the weak-* topology).*

Note: as part of the definition of an "affine" function which can take the value $-\infty$, we require that $s(t\rho_1 + (1-t)\rho_2) = -\infty$ for $\rho_1, \rho_2 \in E^I$ with $s(\rho_1) = -\infty$ and $1 > t > 0$.

Proof. If the restriction of ρ_1 to $C(\Omega_\Lambda)$ is not absolutely continuous with respect to μ_0, then neither is the restriction of $t\rho_1 + (1-t)\rho_2$ for $1 > t > 0$. So in showing $s(\rho)$ is affine, we can assume $\rho_1^{(\Lambda)}$ and $\rho_2^{(\Lambda)}$ both exist. Let f be the function $f(t) = t \ln t$ on $[0,\infty)$. Since f is convex,

$$< tf(\rho_1^{(\Lambda)}) + (1-t) f(\rho_2^{(\Lambda)})>_0 \ \geq \ < f(t\rho_1^{(\Lambda)} + (1-t)\rho_2^{(\Lambda)})>_0 .$$

so S_Λ is concave. On the other hand, using $f(ts) = ts(\ln t + \ln s) = sf(t) + tf(s)$ and $f(t+s) = (t+s)\ln(t+s) \geq f(t) + f(s)$ we obtain

$$S_\Lambda(t\rho_1+(1-t)\rho_2) \leq -<f(t\rho_1^{(\Lambda)})>_0 - <f((1-t)\rho_2^{(\Lambda)})>_0$$

$$= -t<f(\rho_1^{(\Lambda)})>_0 - (1-t)<f(\rho_2^{(\Lambda)})>_0 - f(t)<\rho_1^{(\Lambda)}>_0 - f(1-t)<\rho_2^{(\Lambda)}>_0$$

$$= tS_\Lambda(\rho_1) + (1-t) S_\Lambda(\rho_2) - f(t) - f(1-t) .$$

Taking $\Lambda \to \infty$ (van Hove) we obtain

$$s(t\rho_1 + (1-t)\rho_2) = \lim |\Lambda|^{-1} S_\Lambda(t\rho_1 + (-t)\rho_2) = ts(\rho_1) + (1-t)\, s(\rho_2)$$

so $s(\rho)$ is affine.

The upper semicontinuity of $s(\rho) = \inf 2^{-n\nu} S_{C_{2^n}}(\rho)$ will follow if each S_Λ is upper semicontinuous.

Let ρ be a probability measure on Ω_Λ. If $S_\Lambda(\rho) > -\infty$, we must show that for any $\varepsilon > 0$ there is a weak-* neighborhood U_ε of ρ such that $S_\Lambda(\rho') \leq S_\Lambda(\rho) + \varepsilon$ if $\rho' \in U_\varepsilon$; if $S_\Lambda(\rho) = -\infty$, we must show that for any M there is a weak-* neighborhood U_M of ρ with $S_\Lambda(\rho') \leq -M$ for $\rho' \in U_M$.

Suppose first $S_\Lambda(\rho) > -\infty$. Then $\ln \rho^{(\Lambda)} \in L^1(\rho)$. By suitably cutting it off above and below, and using the fact that $C(\Omega_\Lambda)$ is dense in $L^1(\Omega_\Lambda, \mu_0 + \rho)$, we can find $f \in C(\Omega_\Lambda)$ with $f > 0$, $\rho(|\ln f - \ln \rho^{(\Lambda)}|) < \frac{\varepsilon}{3}$ and $|<f>_0 - 1| < \frac{\varepsilon}{3}$. Then for any state ρ' with $\rho'(\ln f) > \rho(\ln f) - \frac{\varepsilon}{3}$, either $S_\Lambda(\rho') = -\infty$ and there is nothing to prove, or $\rho'^{(\Lambda)}$ exists and

$$\rho'\left(\ln\frac{\rho'^{(\Lambda)}}{f}\right) \geq \rho'\left(1-\frac{f}{\rho'^{(\Lambda)}}\right) = 1-\langle f\rangle_0 > \frac{\varepsilon}{3}$$

so $S_\Lambda(\rho') = -\rho'(\ln \rho'^{(\Lambda)}) \leq \frac{\varepsilon}{3} - \rho'(\ln f) \leq \frac{2\varepsilon}{3} - \rho(\ln f) \leq \varepsilon + S_\Lambda(\rho)$.

Suppose now $\rho^{(\Lambda)}$ exists but $S_\Lambda(\rho) = -\infty$. For $N>0$ let $\rho_N^{(\Lambda)}$ be the function on Ω_Λ

$$\rho_N^{(\Lambda)}(\omega) = \begin{cases} \rho^{(\Lambda)}(\omega) & \text{if} \quad \rho^{(\Lambda)}(\omega) \leq N \\ 1 & \text{if} \quad \rho^{(\Lambda)}(\omega) > N\,. \end{cases}$$

Then $\ln \rho_N^{(\Lambda)} \in L^1(\rho)$. We can find $f \in C(\Omega_\Lambda)$ with $f>0$, $\rho(|\ln f - \ln \rho_N^{(\Lambda)}|) < 1$, and $|\langle f - \rho_N^{(\Lambda)}\rangle_0| < 1$. Now given M we can take N so large that $\rho(\ln \rho_N^{(\Lambda)}) \geq M+4$. Then if $\rho'(\ln f) > \rho(\ln f) - 1$, a similar calculation to that above yields

$$S_\Lambda(\rho') = -\rho'(\ln \rho'^{(\Lambda)}) \leq \langle f\rangle_0 - 1 - \rho'(\ln f) < -M\,.$$

Finally, if $\rho^{(\Lambda)}$ does not exist, there is a compact set $K \subset \Omega_\Lambda$ with $\rho(K) > 0$ and $\mu_0(K) = 0$. Given M, we take $N = \rho(K)^{-1}(M+2)$ and $f \in C(\Omega_\Lambda)$ with $1 \leq f \leq e^N$, $f = e^N$ on K, and $\langle f\rangle_0 \leq 2$. Thus $\rho(\ln f) \geq N\rho(K) = M+2$, and if $\rho'(\ln f) > M+1$ we have $S_\Lambda(\rho') \leq -\rho'(\ln f) + \langle f\rangle_0 - 1 < -M$. ∎

In the quantum lattice system, density matrices will take the place of Radon-Nikodym derivatives. For any state ρ and any finite subset Λ of Z^ν, there is a self-adjoint operator $\rho^{(\Lambda)} \in \mathfrak{A}_\Lambda$ such that $\rho(A) = \mathrm{tr}(\rho^{(\Lambda)}A)$ for $A \in \mathfrak{A}_\Lambda$. Note that $0 \leq \rho^{(\Lambda)} \leq (\dim \mathcal{H}_0)^{|\Lambda|}1$. We then define the *entropy of* ρ *in* Λ by

$$(5) \qquad S_\Lambda(\rho) = -\mathrm{tr}(\rho^{(\Lambda)} \ln \rho^{(\Lambda)})\,.$$

If $\rho^{(\Lambda)}$ is strictly positive, so that $\ln \rho^{(\Lambda)} \in \mathfrak{A}_\Lambda$, this is equal to $-\rho(\ln \rho^{(\Lambda)})$; in any case, $t \ln t$ (defined to be 0 at $t=0$) is a continuous function on $[0,\infty)$ and defines $\rho^{(\Lambda)}\ln\rho^{(\Lambda)}$ by the functional calculus. The more usual definition would use the operator $(\dim\mathcal{H}_\Lambda)\rho^{(\Lambda)}$ instead of $\rho^{(\Lambda)}$ and the trace Tr_Λ instead of tr. As in the classical case, this leads to an entropy $S'_\Lambda(\rho) = S_\Lambda(\rho) + |\Lambda| \ln \dim \mathcal{H}_0$.

The strong subadditivity of entropy in quantum systems is quite difficult to prove: see [37]. We will content ourselves with proving subadditivity.

LEMMA II.2.4. *Let* A, B *be positive nonsingular matrices with* $\mathrm{tr}(A) = \mathrm{tr}(B)$. *Then* $\mathrm{tr}(A \ln A) \geq \mathrm{tr}(A \ln B)$.

Proof. Let $\{u_n\}$ and $\{v_n\}$ be orthonormal bases of eigenvectors for A and B respectively, with $Au_n = a_n u_n$ and $Bv_n = b_n v_n$. Then using the inequality $t \ln t \geq t \ln s + t - s$ for $s,t > 0$ (which is $\lambda \ln \lambda \geq \lambda - 1$ with $\lambda = t/s$), we obtain

$$\langle u_i, A \ln B\, u_i\rangle = \sum_j |\langle u_i,v_j\rangle|^2 a_i \ln b_j \leq \sum_j |\langle u_i,v_j\rangle|^2 (a_i \ln a_i + b_j - a_i)$$

$$= \langle u_i, A \ln A\, u_i\rangle + \langle u_i, (B-A)u_i\rangle$$

so that $\mathrm{tr}(A \ln B) \leq \mathrm{tr}(A \ln A) + \mathrm{tr}(B-A) = \mathrm{tr}(A \ln A)$. ∎

LEMMA II.2.5. *For any state* ρ *on* $\mathfrak{A}$ *and* $\Lambda, \Lambda' \subset Z^\nu$ *finite,*

a) $-\ln \dim \mathcal{H}_\Lambda \leq S(\Lambda) \leq 0$ *(negativity and lower bound).*

b) $\Lambda \subseteq \Lambda' \Rightarrow S(\Lambda') \leq S(\Lambda)$ *(decrease).*

c) $\Lambda \cap \Lambda' = \emptyset \Rightarrow S(\Lambda \cup \Lambda') \leq S(\Lambda) + S(\Lambda')$ *(subadditivity).*

Proof. Part (b) will follow from (a) and (c). For (a) we note that $S(\Lambda) = -\mathrm{tr}(\rho^{(\Lambda)}\ln\rho^{(\Lambda)}) \leq \mathrm{tr}(1-\rho^{(\Lambda)}) = 0$, while $\rho(\ln\rho^{(\Lambda)}) \leq \ln \dim \mathcal{H}_\Lambda$ since $\rho^{(\Lambda)} \leq \dim \mathcal{H}_\Lambda$. For (c), note that $\rho^{(\Lambda)} = \mathrm{tr}_{\Lambda'}\rho^{(\Lambda\cup\Lambda')}$ and $\rho^{(\Lambda')} = \mathrm{tr}_\Lambda \rho^{(\Lambda\cup\Lambda')}$. Let $A = \rho^{(\Lambda\cup\Lambda')}$ and $B = \rho^{(\Lambda)}\rho^{(\Lambda')}$. We can assume A and B are nonsingular (otherwise replace ρ by $(1+\varepsilon)^{-1}(\rho + \varepsilon\,\mathrm{tr})$ and let

$\varepsilon \to 0$). Then by Lemma II.2.4 we have

$$S(\Lambda \cup \Lambda') = -\mathrm{tr}(A \ln A) \leq -\mathrm{tr}(A \ln B) = -\rho(\ln \rho^{(\Lambda)}) - \rho(\ln \rho^{(\Lambda')})$$
$$= S(\Lambda) + S(\Lambda') . \qquad \blacksquare$$

THEOREM II.2.6. *In the quantum lattice system, for* $\rho \in E^I$ *the mean entropy* $s(\rho) = \lim_{a \to \infty} a^{-\nu} S_{C_a}(\rho)$ *exists, where* C_a *is a cube of side* a. *This is an affine upper semicontinuous function on* E^I *(in the weak-* topology), with* $0 \geq s(\rho) \geq -\ln \dim \mathcal{H}_0$.

Proof. The only change from the proof of Theorem II.2.2 is that we have not proven strong subadditivity, so that only a limit of cubes C_a can be considered (it is not hard to show, with the methods of the first two parts of the proof of Theorem II.2.2, that the limit exists as $a \to \infty$, not just over powers of 2). The upper semicontinuity of $s(\rho)$ is easier to obtain here, since S_Λ is clearly continuous. The concavity of S_Λ follows from Lemma II.2.4 by

$$t\, \mathrm{tr}[\rho_1^{(\Lambda)} \ln(t\, \rho_1^{(\Lambda)} + (1-t)\rho_2^{(\Lambda)})] + (1-t)\,\mathrm{tr}[\rho_2^{(\Lambda)} \ln(t\rho_1^{(\Lambda)} + (1-t)\rho_2^{(\Lambda)})]$$
$$\leq t\, \mathrm{tr}(\rho_1^{(\Lambda)} \ln \rho_1^{(\Lambda)}) + (1-t)\,\mathrm{tr}(\rho_2^{(\Lambda)} \ln \rho_2^{(\Lambda)}) .$$

On the other hand, to obtain the inequality

$$S_\Lambda(t\rho_1 + (1-t)\rho_2) \leq t\, S_\Lambda(\rho_1) + (1-t)\, S_\Lambda(\rho_2) - f(t) - f(1-t)$$

we use the fact that for matrices A, B with $0 < A \leq B$ we have $\ln A \leq \ln B$. To prove this, note first that $B^{-1} \leq A^{-1}$ since $0 < B^{-\frac{1}{2}} A B^{-\frac{1}{2}} \leq 1$ implies $1 \leq (B^{-\frac{1}{2}} A B^{-\frac{1}{2}})^{-1} = B^{\frac{1}{2}} A^{-1} B^{\frac{1}{2}}$, and then $\ln B = -\int_0^\infty (x+B)^{-1}\, dx \geq -\int_0^\infty (x+A)^{-1}\, dx = \ln A$. $\blacksquare$

Note that the mean entropies we have defined here differ from the more usual definition by $\ln \dim \mathcal{H}_0$ in the quantum case, and $\ln |\Omega_0|$ in the classical (discrete-spin) case.

II.3. *The variational principle*

Equilibrium states of finite systems can be characterized by means of a variational principle, and this can be used to give a criterion for invariant equilibrium states of the infinite system. We will begin with a classical system in the finite set $\Lambda \subset \mathbf{Z}^{\nu}$ with Hamiltonian $H_\Lambda \in C(\Omega_\Lambda)$. The equilibrium state ρ_1 of the system is given by $\langle A\rangle_1 = Z_\Lambda^{-1}\langle e^{-H_\Lambda}A\rangle_0$ where $Z_\Lambda = \langle e^{-H_\Lambda}\rangle_0$. Thus the pressure, mean energy and entropy are given by

$$P_\Lambda = |\Lambda|^{-1} \ln Z_\Lambda$$

$$\langle H_\Lambda\rangle_1 = Z_\Lambda^{-1}\langle H_\Lambda e^{-H_\Lambda}\rangle_0$$

$$\begin{aligned} S_\Lambda(\rho_1) &= -\langle Z_\Lambda^{-1} e^{-H_\Lambda} \ln(Z_\Lambda^{-1} e^{-H_\Lambda})\rangle_0 \\ &= \ln Z_\Lambda + Z_\Lambda^{-1}\langle H_\Lambda e^{-H_\Lambda}\rangle_0 = |\Lambda| P_\Lambda + \langle H_\Lambda\rangle_1 \,. \end{aligned}$$

On the other hand, for any state ρ with $S_\Lambda(\rho) > -\infty$ we have using Jensen's Inequality (Lemma I.2.1)

$$\begin{aligned} S_\Lambda(\rho) - \rho(H_\Lambda) &= \rho(-\ln \rho^{(\Lambda)} - H_\Lambda) \le \ln \rho((\rho^{(\Lambda)})^{-1} e^{-H_\Lambda}) \\ &= \ln \langle e^{-H_\Lambda}\rangle_0 = |\Lambda| P_\Lambda \,. \end{aligned}$$

It can easily be shown that in order for equality to hold in Jensen's Inequality $e^{\int f\, d\mu} \le \int e^f d\mu$ the function f must be constant almost everywhere. Therefore the equilibrium state ρ_1 is the only state ρ of the finite system for which $S_\Lambda(\rho) - \rho(H_\Lambda)$ attains its maximum $|\Lambda| P_\Lambda$.

The quantum system in Λ with Hamiltonian $H_\Lambda \in \mathfrak{A}_\Lambda$ can be dealt with in a similar way. Again we can calculate $S_\Lambda(\rho_1) - \rho_1(H_\Lambda) = |\Lambda| P_\Lambda$ for the equilibrium state $\rho_1(A) = Z_\Lambda^{-1}\,\mathrm{tr}(e^{-H_\Lambda}A)$. To obtain the variational inequality $S_\Lambda(\rho) - \rho(H_\Lambda) \le |\Lambda| P_\Lambda$ for an arbitrary state ρ, we take an orthonormal basis of $\mathcal{H}_\Lambda$ consisting of eigenvectors u_j of $\rho^{(\Lambda)}$, with

$\rho^{(\Lambda)}u_j = c_j u_j$. Note that $c_j \geq 0$ and $\sum_{j=1}^{d} c_j = d$ where d is the dimension of $\mathcal{H}_\Lambda$. Then by Lemmas I.2.1 and I.3.1

$$e^{S_\Lambda(\rho)-\rho(H_\Lambda)} = e^{tr(\rho^{(\Lambda)}(-\ln \rho^{(\Lambda)} - H_\Lambda))}$$

$$= \exp d^{-1} \sum_{c_j \neq 0} c_j(-\ln c_j - \langle u_j, H_\Lambda u_j\rangle)$$

$$\leq d^{-1} \sum_{c_j \neq 0} c_j e^{-\ln c_j - \langle u_j, H_\Lambda u_j\rangle} = d^{-1} \sum_{c_j \neq 0} e^{-\langle u_j, H_\Lambda u_j\rangle}$$

$$\leq d^{-1} \sum_{j=1}^{d} \langle u_j, e^{-H_\Lambda} u_j\rangle = \operatorname{tr} e^{-H_\Lambda} = e^{|\Lambda| P_\Lambda}.$$

In order for equality to hold here we must have $e^{-\langle u_j, H_\Lambda u_j\rangle} = \langle u_j, e^{-H_\Lambda} u_j\rangle$ for each j. By considering the spectral representation of H_Λ it is not difficult to see that u_j must then be an eigenvector for H_Λ. As in the classical case we then find that $S_\Lambda(\rho) - \rho(H_\Lambda) = |\Lambda| P_\Lambda$ only for the equilibrium state ρ_1. We summarize these results as a lemma:

LEMMA II.3.1. *For any state* ρ *of the classical or quantum system in the finite set* $\Lambda \subset Z^\nu$ *with Hamiltonian* H_Λ,

$$S_\Lambda(\rho) - \rho(H_\Lambda) \leq |\Lambda| P_\Lambda. \tag{6}$$

Equality holds in (6) *if and only if* ρ *is the thermal equilibrium state* ρ_1.

Now for translation-invariant states of the infinite system with an interaction Φ we can divide each term of (6) by $|\Lambda|$ and take limits. Note that for any $\rho \in E^I$ and $\Phi \in \mathcal{B}$, as $\Lambda \to \infty$ (van Hove) we have

$$|\Lambda|^{-1}\rho(H_\Lambda^\Phi) = \sum_{X\subset\Lambda} |\Lambda|^{-1}\rho(\Phi(X)) = \sum_{Y\ni 0} |Y|^{-1} M_Y^-(\Lambda)|\Lambda|^{-1}\rho(\Phi(Y))$$

$$\to \sum_{Y\ni 0} |Y|^{-1}\rho(\Phi(Y)) = \rho(A_\Phi)\,.$$

THEOREM II.3.2 (see [24]). *For any* $\Phi \in \mathcal{B}$,

(7) $$P(\Phi) = \sup\{s(\rho)-\rho(A_\Phi) : \rho \in E^I\}\,.$$

Proof. From the inequality (6) for the finite systems we obtain $P(\Phi) \geq s(\rho)-\rho(A_\Phi)$ for all $\rho \in E^I$. To show that $P(\Phi)$ is actually the least upper bound, we will use the equilibrium states of finite systems, for which equality holds in (6). Thus we let C_a be a cube of side a, and ρ_1 the thermal equilibrium state for the system in C_a with Hamiltonian $H_{C_a}^\Phi$. A state of the infinite system is obtained by letting the cubes $C_a + aj$, $j \in Z^\nu$, be independent and taking the appropriate translate of ρ_1 in each (i.e., we take the state ρ' as the product of the translates of ρ_1 operating in each cube $C_a + aj$). It is easy to see that $s(\rho') = a^{-\nu} S_{C_a}(\rho_1) = P_{C_a}(\Phi) + a^{-\nu}\rho_1(H_{C_a}^\Phi)$. The state ρ' is periodic but not translation-invariant; to obtain an invariant state we must average over translations. Thus we let $\rho = a^{-\nu}\sum_{j\in C_a} \rho'\circ\tau_j$. Since the function s is affine (on periodic states as well as invariant ones) we have $s(\rho) = s(\rho')$. For a fixed finite subset X of Z^ν containing 0 we have

$$\left| a^\nu \rho(\Phi(X)) - \rho_1\left(\sum_{i+X\subset C_a} \Phi(i+X)\right)\right| = \left|\sum_{\substack{i\in C_a\\ i+X\not\subset C_a}} \rho'\circ\tau_i(\Phi(X))\right|$$

$$\leq (a^\nu - M_X^-(C_a))\,\|\Phi(X)\|\,.$$

Thus for any $\varepsilon > 0$ we can take a sufficiently large that $|P(\Phi) - P_{C_a}(\Phi)|$ and $|\rho(A_\Phi) - \rho_1(H^\Phi_{C_a})|$ are both less than $\varepsilon/2$, so that we obtain $s(\rho) - \rho(A_\Phi) \geq P(\Phi) - \varepsilon$. ■

COROLLARY II.3.3. *Any invariant state* ρ *with* $P(\Phi) = s(\rho) - \rho(A_\Phi)$ *is an invariant equilibrium state (in the sense of Section* II.1) *for* Φ.

Proof. For any $\Psi \in \mathcal{B}$, $P(\Phi+\Psi) \geq s(\rho) - \rho(A_\Phi) - \rho(A_\Psi) = P(\Phi) - \rho(A_\Psi)$. ■

Our next goal is to prove the converse of this corollary, i.e., that invariant equilibrium states satisfy the variational equality $P(\Phi) = s(\rho) - \rho(A_\Phi)$. Note that if ρ is an invariant equilibrium state for $\Phi \in \mathcal{B}$, i.e., the functional $a(\Psi) = -\rho(A_\Psi)$ on $\mathcal{B}$ is tangent to P at Φ, we have $P(\Phi) + \rho(A_\Phi) = \inf\{P(\Psi) + \rho(A_\Psi): \Psi \in \mathcal{B}\}$. Thus the desired result is a consequence of the "inverse variational principle" $s(\rho) = \inf\{P(\Phi) + \rho(A_\Psi): \Psi \in \mathcal{B}\}$. The original method of proof (see [17]) is very indirect: first it is shown that if the tangent functional to P at Φ is unique, it determines an invariant state by (6) which satisfies the variational equality; then arbitrary tangent functionals are approximated weakly by convex combinations of unique tangent functionals at interactions approaching the given interaction (using the fact that a dense set of interactions have unique tangent functionals, see [8]); and then the "inverse variational principle" is derived, using the fact that invariant equilibrium states are weak-* dense in E^I. We will provide two direct proofs of the "inverse variational principle." The first, due to Ruelle (unpublished), shows that the two variational principles are related by "duality." The second proof actually constructs a sequence of interactions Ψ_n for which $P(\Psi_n) + \rho(A_{\Psi_n})$ tends to $s(\rho)$.

THEOREM II.3.4. *For an invariant state* ρ *of the classical or quantum lattice system,* $s(\rho) = \inf\{P(\Psi) + \rho(A_\Psi): \Psi \in \mathcal{B}\}$.

First Proof (Ruelle). Suppose $r > s(\rho)$. The subgraph $\{(\rho',t): \rho' \in E^I, t \leq s(\rho')\}$ of s is a closed convex set in $M(\Omega) \oplus R$ (or $\mathfrak{A}^* \oplus R$ for the quantum system) with the weak-* topology. By assumption it does not contain the point (ρ, r). Therefore the two can be separated by a hyperplane: there are $\lambda \in R$ and $A \in C(\Omega)$ or $\mathfrak{A}$ (corresponding to a weak-* continuous linear functional on $M(\Omega)$ or $\mathfrak{A}^*$) with

$$\rho(A) + \lambda r > \varepsilon + \sup\{\rho'(A) + \lambda s(\rho'): \rho' \in E^I, s(\rho) > -\infty\}$$

for some $\varepsilon > 0$. It is clear, except perhaps in the classical non-discrete case, that $\lambda > 0$; by the Hahn-Banach theorem, showing $\lambda > 0$ there is equivalent to showing that states of finite mean entropy are weak-* dense in E^I. This can be done fairly easily (and we omit the proof). We can take the normalization $\lambda = 1$. By approximating A by a member B of a local algebra $C(\Omega_\Lambda)$ or $\mathfrak{A}_\Lambda$, we obtain the interaction $\Psi = -\Psi_B^\Lambda$ with

$$\rho(-A_\Psi) + r = \rho(B) + r > \sup\{\rho'(B) + s(\rho'): \rho' \in E^I, s(\rho') > -\infty\} = P(\Psi)$$

$$\text{and } r > P(\Psi) + \rho(A_\Psi). \qquad \blacksquare$$

Second Proof (classical discrete case). We know $s(\rho) \leq P(\Psi) + \rho(A_\Psi)$. We may assume the support of the measure ρ is all of Ω: otherwise replace ρ by $\rho_t = t\rho + (1-t)\mu_0$, $0 \leq t < 1$. Since both $s(\rho_t)$ and $P(\Psi) + \rho_t(A_\Psi)$ are affine functions of t, if the theorem is true for ρ_t it must also be true for ρ.

Let C be a ν-dimensional cube in Z^ν. We want to find an interaction Φ so that the restriction $\rho^{(C)} d\mu_0$ of ρ to $C(\Omega_C)$ will look as much as possible like the equilibrium state for the finite system. By assumption $\rho^{(C)}$ is strictly positive, and since we are dealing with the discrete case there are no continuity problems. So we can take $\Phi = \Psi_F^C$ with $F = -|C|^{-1} \ln \rho^{(C)}$. Thus we have $\rho(A_\Phi) = -|C|^{-1} \rho(\ln \rho^{(C)}) = |C|^{-1} S_C(\rho)$.

To compute $P(\Phi)$ it is convenient to use periodic boundary conditions. Let $T_n = (\mathbb{Z}/n\mathbb{Z})^\nu$ where n is a multiple of the side of C. We can write

$$H^{\Phi}_{T_n} = |C|^{-1} \sum_{j \in C} G_j \tag{8}$$

with

$$G_j = |C| \sum_{\substack{i \in T_n \\ i \equiv j \bmod C}} \Phi(C+i) = - \sum_{\substack{i \in T_n \\ i \equiv j \bmod C}} \ln \rho^{(C+i)} .$$

By convexity of the function $f \mapsto \ln \langle e^{-f} \rangle_0$ (as in Theorem I.2.3) we have

$$P_{T_n}(\Phi) = n^{-\nu} \ln \langle \exp(-H^{\Phi}_{T_n}) \rangle_0 \leq |C|^{-1} n^{-\nu} \sum_{j \in C} \ln \langle \exp(-G_j) \rangle_0 .$$

Now for fixed j the sets $C+i$ for $i \equiv j \bmod C$ cover T_n by $|C|^{-1} n^\nu$ disjoint cubes. Therefore we have

$$\exp(-G_j) = \prod_{\substack{i \equiv j \\ \bmod C}} \rho^{(C+i)}$$

$$\langle \exp(-G_j) \rangle_0 = \prod_{\substack{i \equiv j \\ \bmod C}} \langle \rho^{(C+i)} \rangle_0 = 1$$

so $P_{T_n}(\Phi) \leq 0$, and $P(\Phi) + \rho(A_\Phi) \leq |C|^{-1} S_C(\rho)$. Taking $C \to \infty$ we obtain $\inf\{P(\Psi) + \rho(A_\Psi) : \Psi \in \mathcal{B}\} \leq s(\rho)$.

The proof of the quantum case is essentially the same as the above proof of the classical discrete case. The $\rho^{(C)}$ are now density matrices, and the normalized trace tr replaces μ_0. Lemma I.3.3 is used to show $P_{T_n}(\Phi) \leq |C|^{-1} n^{-\nu} \sum_{j \in C} \ln \operatorname{tr} \exp(-G_j)$.

Proof of general classical case. There are two difficulties which arise when Ω_0 is not discrete. First, the restriction of ρ to $C(\Omega_C)$ may not be absolutely continuous with respect to μ_0. Second, if the Radon-Nikodym derivative $\rho^{(C)}$ exists it may not be continuous.

If the restriction of ρ to $C(\Omega_C)$ is not absolutely continuous with respect to μ_0, there is a compact set $E \subset \Omega_C$ with $\rho(E) > 0$ and $\mu_0(E) = 0$. Given $N > 0$ we can find a function $A \in C(\Omega_C)$ with $-N \leq A \leq 0$, $A = -N$ on E, and $\langle e^{-|C|A} \rangle_0 \leq e$. Consider the interaction $\Psi = \Psi_A^C$. Then $\rho(A_\Psi) = \rho(A) \leq -N\rho(E)$. For estimating the pressure we again use T_n where n is a multiple of the side of the cube C. We have

$$H_{T_n}^{\Psi} = |C|^{-1} \sum_{j \in C} G_j$$

with

$$G_j = |C| \sum_{\substack{i \in T_n \\ i \equiv j \bmod C}} \tau_i A$$

so that

$$P_{T_n}(\Psi) \leq |C|^{-1} n^{-\nu} \sum_{j \in C} \ln \langle \exp(-G_j) \rangle_0$$

but

$$\langle \exp(-G_j) \rangle_0 = \prod_{i \equiv j \bmod C} \langle \exp(-|C| \tau_i A) \rangle_0 \leq e^{|C|^{-1} n^{\nu}}$$

so

$$P_{T_n}(\Psi) \leq |C|^{-1} .$$

Thus

$$P(\Psi) + \rho(A_\Psi) \leq |C|^{-1} - N\rho(E) \to -\infty = s(\rho) \text{ as } N \to \infty .$$

In the second situation, where $\rho^{(C)}$ is not continuous, we must approximate it by a continuous function. By the same argument that said we could assume the support of ρ was all of Ω, we can assume that $\rho^{(C)} \geq \varepsilon$ for some $\varepsilon > 0$. There are again two cases, depending on whether $S_C(\rho) = -\infty$.

If $S_C(\rho) = -\rho(\ln \rho^{(C)}) = -\infty$, then for any M there is N with

$$\rho(\ln \inf(\rho^{(C)}, N)) > |C|\, M .$$

We can take $\Psi = \Psi_A^C$, choosing $A \in C(\Omega_C)$ with

$$\rho(|A + |C|^{-1} \ln \inf(\rho^{(C)}, N)|) < 1$$

and

$$\mu_0(|e^{-|C|A} - \inf(\rho^{(C)}, N)|) < 1 .$$

Then

$$\rho(A_\Psi) = \rho(A) \leq 1 - |C|^{-1} \rho(\ln \inf(\rho^{(C)}, N)) < 1 - M .$$

Again we estimate $P(\Psi)$ using periodic boundary conditions. We have

$$< e^{-|C|A} >_0 \; \leq \; < \rho^{(C)} >_0 + 1 = 2$$

so

$$\ln < \exp(-G_j) >_0 \; \leq |C|^{-1} n^\nu \ln 2$$

and

$$P(\Psi) \leq |C|^{-1} \ln 2 .$$

Taking $M \to \infty$ gives us $\inf\{P(\Psi) + \rho(A_\Psi) : \Psi \in \mathcal{B}\} = -\infty$.

If $S_C(\rho) > -\infty$ and $\delta > 0$ is given, we can find $A \in C(\Omega_C)$ with

$$\rho(|A + |C|^{-1} \ln \rho^{(C)}|) < \delta$$

and

$$\mu_0(|e^{-|C|A} - \rho^{(C)}|) < \delta .$$

Our estimates then become

$$\rho(A_\Psi) \leq \delta + |C|^{-1} S_C(\rho)$$

and

$$<e^{-|C|A}>_0 \leq 1+\delta$$

yielding $P(\Psi) \leq |C|^{-1} \ln(1+\delta)$. Thus as $\delta \to 0$ we obtain $\inf\{P(\Psi) + \rho(A_\Psi) : \Psi \in \mathcal{B}\} \leq |C|^{-1} S_C(\rho)$. To complete the proof we take $C \to \infty$.

COROLLARY II.3.5. *If* ρ *is an invariant equilibrium state for* $\Phi \in \mathcal{B}$, *then* $P(\Phi) = s(\rho) - \rho(A_\Phi)$.

Proof. By the remarks before Theorem II.3.4. ∎

III. DLR EQUATIONS AND KMS CONDITIONS

III.1. *The DLR equations*

In Chapter II we defined invariant equilibrium states in terms of tangent functionals to the pressure, and showed that this was equivalent to the formulation in terms of a Variational Principle. An alternative approach is to define (not necessarily invariant) equilibrium states, using the Dobrushin-Lanford-Ruelle (DLR) equations [18] for classical systems and the Kubo-Martin-Schwinger (KMS) boundary conditions [14] for quantum systems. The DLR equations are motivated by the consideration of relative probabilities of configurations in a finite system with fixed boundary conditions. The KMS conditions are rather more subtle, and require consideration of the time evolution of the quantum system. We begin with the classical system and the DLR equations.

For motivational purposes, consider first the case where Ω_0 is discrete and the interaction Φ involves only nearest neighbors ($\Phi(X) = 0$ if $|X| > 2$ or $\operatorname{diam}(X) > 1$). Let Λ be a finite subset of Z^ν, and let $\partial\Lambda$ be the set of sites in the complement Λ^c of Λ which are nearest neighbors of sites in Λ. Consider two configurations $\omega_1, \omega_2 \in \Omega_\Lambda$ and a fixed "boundary condition" configuration $\tau \in \Omega_{\partial\Lambda}$. The Gibbs equilibrium state ρ_τ for the finite system in Λ with Hamiltonian ${}_\tau H_\Lambda^\Phi$ (as in formula (18) of Section I.2) assigns the configurations ω_1 and ω_2 the probabilities $\rho_\tau(\{\omega_1\})$ and $\rho_\tau(\{\omega_2\})$ related by

$$\rho_\tau(\{\omega_1\}) = \exp({}_\tau H_\Lambda^\Phi(\omega_2) - {}_\tau H_\Lambda^\Phi(\omega_1))\rho_\tau(\{\omega_2\}) . \tag{1}$$

In the equilibrium state for the system in a larger finite region Λ' containing both Λ and $\partial\Lambda$, the probability that the configuration in Λ is

ω_1 or ω_2 will depend on the probabilities of various configurations τ occurring in $\partial\Lambda$; however, the conditional probabilities for ω_1 and ω_2 given the configuration τ in $\partial\Lambda$ will still be related as in (1). This should be expected to remain true if the equilibrium state for the finite system in Λ' were replaced by an "equilibrium state" for the infinite system. Thus one might define an equilibrium state for Φ to be a state ρ such that for all Λ, ω_1, ω_2 and τ as above, the conditional probabilities $\rho(\{\omega_i\}|\tau)$ for configurations ω_i in Λ given the configuration τ in $\partial\Lambda$ are related by

$$\rho(\{\omega_1\}|\tau) = \exp({}_\tau H^\Phi_\Lambda(\omega_2) - {}_\tau H^\Phi_\Lambda(\omega_1))\rho(\{\omega_2\}|\tau) .$$

Of course, we want a definition of equilibrium states for more general interactions than nearest-neighbor ones. For finite-range interactions we can simply enlarge the "boundary" $\partial\Lambda$ to include all sites $i \in \Lambda^c$ with $\Phi(X) \neq 0$ for some X containing i and intersecting Λ. For infinite-range interactions $\partial\Lambda$ might become all of Λ^c; some care is then necessary in interpreting the "conditional probabilities" because the probability of a given configuration occurring for an infinite region is likely to be zero. We will also want to remove the restriction to discrete Ω_0. This will involve replacing conditional probabilities by "conditional probability densities."

We consider a fixed interaction $\Phi \in \tilde{\mathcal{B}}$ (for the classical system, without requiring Ω_0 to be discrete), and a finite region $\Lambda \subset Z^\nu$. Given a state ρ (of the infinite system), we want to define associated measures $\rho(s, d\tau)$ on Ω_{Λ^c} for each $s \in \Omega_\Lambda$, so that

$$\rho(f) = \int_{\Omega_\Lambda} \int_{\Omega_{\Lambda^c}} f(s \times \tau)\rho(s, d\tau)\mu_0(ds) \quad \text{for } f \in C(\Omega) . \tag{2}$$

Taking f of the form $g \otimes h$ with $g \in C(\Omega_\Lambda)$ and $h \in C(\Omega_{\Lambda^c})$, we see that this means

$$\rho(g \otimes h) = \int_{\Omega_\Lambda} g(s)\rho(s,h)\mu_0(ds)$$

i.e., $\rho(s,h)$ is the Radon-Nikodym derivative $\rho(ds \otimes h)/\mu_0(ds)$. If Ω_0 is discrete there is no problem here: $\rho(s,h) = (\mu_0(\chi_s))^{-1}\rho(\chi_s \otimes h)$ where χ_s is the function on Ω_Λ defined by $\chi_s(s') = 1$ if $s' = s$, 0 otherwise. However, in the general case $\rho(ds \otimes h)$ may not be absolutely continuous with respect to μ_0, and even if it is, $\rho(s,h)$ is not uniquely defined for all s, but only up to sets of μ_0-measure zero in Ω_Λ. The DLR equations will require that the measures $\rho(s, d\tau)$ exist and are chosen in such a way that $\rho(\cdot, h)$ is a continuous function on Ω_Λ; the requirement of this continuity then leads to a unique choice of the $\rho(s, d\tau)$.

It will be convenient to define a function $W \in C(\Omega_\Lambda \times \Omega)$ by

$$W(s_1, s_2 \times \tau) = {}_\tau H_\Lambda^\Phi(s_1) - {}_\tau H_\Lambda^\Phi(s_2) = \sum_{X \cap \Lambda \neq \emptyset} [\Phi(X)(s_1 \times \tau) - \Phi(X)(s_2 \times \tau)]$$

for $s_1, s_2 \in \Omega_\Lambda$, $\tau \in \Omega_{\Lambda^c}$, giving the difference in energy between configurations s_1 and s_2 in Λ with boundary condition $\tau \in \Omega_{\Lambda^c}$. Since the sum converges absolutely, W is continuous, and $\|W\|_\infty \leq 2\|\Phi\|_{\sim}$. We will say that the state ρ satisfies the *DLR equations* for the interaction $\Phi \in \tilde{\mathfrak{B}}$ if for all finite $\Lambda \subset \mathbf{Z}^\nu$ and $s \in \Omega_\Lambda$, the measures $\rho(s, d\tau)$ exist and can be chosen to satisfy

(3) $$\rho(s_1, d\tau) = e^{-W(s_1, s_2 \times \tau)}\rho(s_2, d\tau) \text{ for all } s_1, s_2 \in \Omega_\Lambda .$$

Note that this implies that $\rho(s,h) = \int_{\Omega_{\Lambda^c}} h(\tau) e^{-W(s, s_0 \times \tau)} \rho(s_0, d\tau)$ (for any fixed $s_0 \in \Omega_\Lambda$), which is a continuous function of s.

If the state ρ satisfies the DLR equations for Φ, the restriction of ρ to $C(\Omega_\Lambda)$ is $\rho^{(\Lambda)}(s)\,\mu_0(ds)$ where

$$\rho^{(\Lambda)}(s) = \rho(s,1) = \int_{\Omega_{\Lambda^c}} e^{-W(s,s_0\times\tau)} \rho(s_0, d\tau)$$

for arbitrary fixed $s_0 \in \Omega_\Lambda$. We can define a probability measure μ on Ω_{Λ^c} by $\mu(d\tau) = \langle \exp(-{}_\tau H_\Lambda^\Phi)\rangle_0 \exp({}_\tau H_\Lambda^\Phi(s_0))\rho(s_0, d\tau)$ so that

$$\rho^{(\Lambda)}(s) = \int_{\Omega_{\Lambda^c}} \langle \exp(-{}_\tau H_\Lambda^\Phi)\rangle_0^{-1} \exp(-{}_\tau H_\Lambda^\Phi(s))\mu(d\tau) .$$

It is clear from the above formula that μ is a probability measure, since $\langle \rho^{(\Lambda)}\rangle_0 = 1$. This means that the finite-volume restrictions of ρ are averages under the probability measure μ of the finite-volume Gibbs equilibrium states $\rho_\tau(A) = \langle \exp(-{}_\tau H_\Lambda^\Phi)\rangle_0^{-1} \langle \exp(-{}_\tau H_\Lambda^\Phi) A\rangle_0$ for boundary conditions $\tau \in \Omega_{\Lambda^c}$.

LEMMA III.1.1 (see [18]). *If* ρ *is an invariant state satisfying the DLR equations* (3) *for* $\Phi \in \tilde{\mathcal{B}}$, *then* ρ *is an invariant equilibrium state for* Φ.

Proof. We will show that ρ satisfies the Variational Principle $P(\Phi) = s(\rho) - \rho(A_\Phi)$ (see Section II.3). Given $\varepsilon > 0$, by the estimates of Section I.2 we can take Λ sufficiently large that $\|{}_\tau H_\Lambda^\Phi - H_\Lambda^\Phi\|_\infty \le \varepsilon|\Lambda|$ for all $\tau \in \Omega_{\Lambda^c}$. Let μ be defined as above, so that the restriction of ρ to $C(\Omega_\Lambda)$ is the average under μ of the equilibrium states ρ_τ for boundary conditions τ. By Lemma II.3.1 we know $S_\Lambda(\rho_\tau) = |\Lambda|{}_\tau P_\Lambda(\Phi) + \rho_\tau({}_\tau H_\Lambda^\Phi)$ for each $\tau \in \Omega_{\Lambda^c}$. Thus by the concavity of the function $-t \ln t$,

$$S_\Lambda(\rho) \ge \int_{\Omega_{\Lambda^c}} S_\Lambda(\rho_\tau)\mu(d\tau) = \int_{\Omega_{\Lambda^c}} (|\Lambda|{}_\tau P_\Lambda(\Phi) + \rho_\tau({}_\tau H_\Lambda^\Phi))\mu(d\tau)$$

$$\ge |\Lambda| P_\Lambda(\Phi) + \rho(H_\Lambda^\Phi) - 2\varepsilon|\Lambda| .$$

As $\Lambda \to \infty$ (van Hove) we have $|\Lambda|^{-1} S_\Lambda(\rho) \to s(\rho)$, $P_\Lambda(\Phi) \to P(\Phi)$ and $|\Lambda|^{-1} \rho(H_\Lambda^\Phi) \to \rho(A_\Phi)$. ∎

III.2. *Invariant equilibrium states and the DLR equations*

If the DLR equations are used to define equilibrium states, we should justify our definition of "invariant equilibrium states" in terms of tangent functionals by showing that for an invariant state, the properties of being an equilibrium (DLR) state and an invariant equilibrium state for an interaction $\Phi \in \tilde{\mathcal{B}}$ are equivalent. Lemma III.1.1 proved the implication in one direction: invariant DLR states are invariant equilibrium states. Now the usual proof [18] of the converse implication uses very indirect methods, similar to those we were able to avoid in Chapter II in proving that tangent functionals to the pressure determine invariant states satisfying the Variational Principle. In this section we provide a more direct proof. The same method we use here will be used in Section III.3 to show that invariant equilibrium states (for interactions in the appropriate Banach space) satisfy the KMS conditions.

Suppose we want to show that every invariant equilibrium state for interaction Φ satisfies a certain set of conditions of the form $\rho(F) = 0$ with $F \in C(\Omega)$ real. If $F \in C(\Omega_\Lambda)$ for some finite Λ, we can consider the interaction Ψ_F^Λ. Then the desired result is equivalent to saying

$$\frac{d}{d\lambda} P(\Phi + \lambda \Psi_F^\Lambda)|_{\lambda=0} = 0 .$$

In general, there will not be such a neat formulation. However, we can approximate F by finite-range functions: given $\varepsilon > 0$, let $\tilde{F} \in C(\Omega_{\tilde{\Lambda}})$ for some finite $\tilde{\Lambda}$ with $\|F - \tilde{F}\|_\infty < \varepsilon$. Then if we can show that

$$P(\Phi + \lambda \Psi_{\tilde{F}}^{\tilde{\Lambda}}) \leq P(\Phi) + 2\varepsilon|\lambda| \quad \text{for } |\lambda| \text{ sufficiently small} \tag{4}$$

we will have $|\rho(F)| \leq 3\varepsilon$ for every invariant equilibrium state ρ for Φ.

As ε was arbitrary, this means $\rho(F) = 0$. To ease the clutter of notation, we will denote $\Psi^{\tilde{\Lambda}}_{\tilde{F}}$ by Ψ.

To obtain our desired estimate (4) on $P(\Phi + \lambda\Psi)$, we will approximate the pressure using periodic boundary conditions. Note that if T_n is large enough to contain a copy of $\tilde{\Lambda}$

$$\frac{d}{d\lambda} P_{T_n}(\Phi+\lambda\Psi) = \langle \exp(-H^{\Phi+\lambda\Psi}_{T_n})\rangle_0^{-1} \langle -\tilde{F} \exp(-H^{\Phi+\lambda\Psi}_{T_n})\rangle_0$$

is the expectation of $-\tilde{F}$ in the Gibbs equilibrium state for interaction $\Phi + \lambda\Psi$ in T_n (with periodic boundary conditions). We will denote this equilibrium state by $\langle\cdot\rangle_{n,\lambda}$. Now suppose there are functions $F_{n,\lambda} \in C(\Omega_{T_n})$ with $\langle F_{n,\lambda}\rangle_{n,\lambda} = 0$ and $\eta > 0$ such that for all sufficiently large n, $\|F - F_{n,\lambda}\|_\infty \le \varepsilon$ for $|\lambda| < \eta$. Then for $|\lambda| < \eta$, $|\langle\tilde{F}\rangle_{n,\lambda}| \le 2\varepsilon$, and so

$$P_{T_n}(\Phi+\lambda\Psi) \le P_{T_n}(\Phi) + 2\varepsilon|\lambda| \quad \text{for } |\lambda| < \eta .$$

The proof is then completed by taking $n \to \infty$.

To sum up, there are three things we must do to show all invariant equilibrium states for Φ (with $\Phi \in \tilde{\mathcal{B}}$) satisfy a given set of conditions: i) formulate the conditions in the form $\rho(F) = 0$ for certain F, ii) find $F_{n,\lambda}$ with $\langle F_{n,\lambda}\rangle_{n,\lambda} = 0$, and iii) find $\eta > 0$ so that for all sufficiently large n, $\|F - F_{n,\lambda}\|_\infty \le \varepsilon$ for $|\lambda| < \eta$.

We must now try to formulate the DLR equations in the form $\rho(F) = 0$. This will be somewhat simpler in the discrete case, so we will treat this case first. Recall that for $s \in \Omega_X$ the function $\chi_s \in C(\Omega_X)$ was defined by $\chi_s(s') = 1$ if $s = s'$, 0 otherwise. Since for any Λ the linear combinations of functions χ_s for configurations s in finite sets $X \subset \Lambda^c$ are dense in $C(\Omega_{\Lambda^c})$, it suffices to check the DLR equations by letting the measures act on these functions. Let $\Lambda' = \Lambda \cup X$, $s'_i = s_i \times s$. Then the equations

$$\rho[\chi_{s'_1} - e^{-W(s'_1, s'_2 \times \cdot)} \chi_{s'_2}] = 0 \quad \text{for all } \Lambda', s'_1, s'_2$$

are easily seen to be equivalent to the DLR equations (3). Thus we can take $F = \chi_{s'_1} - \exp(-W(s'_1, s'_2 \times \cdot))\chi_{s'_2}$. This function is then approximated by $\tilde{F} \in C(\Omega_{\tilde{\Lambda}})$ for some finite $\tilde{\Lambda}$ containing Λ'.

Now we must find our functions $F_{n,\lambda}$. We identify Λ with some copy of itself contained in T_n (which is assumed large enough to contain it). Then we take

$$F_{n,\lambda} = \chi_{s'_1} - \exp(-W_\lambda)\chi_{s'_2}$$

where W_λ is the function defined on $\Omega_{T_n \setminus \Lambda'}$ by

$$\begin{aligned} W_\lambda(\tau) &= H_{T_n}^{\Phi+\lambda\Psi_{\tilde{F}}}(s'_1 \times \tau) - H_{T_n}^{\Phi+\lambda\Psi_{\tilde{F}}}(s'_2 \times \tau) \\ (5) \qquad &= H_{T_n}^{\Phi}(s'_1 \times \tau) - H_{T_n}^{\Phi}(s'_2 \times \tau) + \lambda \sum_{(i+\tilde{\Lambda}) \cap \Lambda' \neq \emptyset} [\tau_i \tilde{F}(s'_1 \times \tau) - \tau_i \tilde{F}(s'_2 \times \tau)]. \end{aligned}$$

Thus clearly $\langle F_{n,\lambda} \rangle_{n,\lambda} = 0$. Now (this time identifying T_n with a cube C_n containing $\tilde{\Lambda}$) $H_{T_n}^{\Phi}(s'_1 \times \cdot) - H_{T_n}^{\Phi}(s'_2 \times \cdot)$ tends uniformly to $W(s'_1, s'_2 \times \cdot)$ as $n \to \infty$, while the terms involving λ on the right-hand side of (5) are bounded by $2|\tilde{\Lambda}|\,|\Lambda'|\,\|\tilde{F}\|_\infty |\lambda|$. So for $|\lambda|$ sufficiently small and n sufficiently large,

$$\|F - F_{n,\lambda}\|_\infty \leq e^{\|W\|_\infty}(e^{\|W(s'_1, s'_2 \times \cdot) - W_\lambda\|_\infty} - 1) \leq \varepsilon .$$

This is all we need in the discrete case. But now we must proceed to the general case. Here we can not keep s'_1 and s'_2 fixed, since single points of $\Omega_{\Lambda'}$ will in general have μ_0-measure zero. Instead, we will consider the equations

(6) $$\rho(f - T_\Lambda f) = 0 \text{ for all } f \in C(\Omega),\ \Lambda \subset Z^\nu \text{ finite}$$

where

(7) $$T_\Lambda f(s\times\tau) = \int_{\Omega_\Lambda} e^{W(s,s'\times\tau)} f(s'\times\tau)\mu_0(ds') \text{ for } s \in \Omega_\Lambda,\ \tau \in \Omega_{\Lambda^c}.$$

We claim that (6) is equivalent to the DLR equations (3). If (6) holds, we have

$$\rho(f) = \int_{\Omega_\Lambda}\int_{\Omega} e^{W(s,s'\times\tau)} f(s'\times\tau)\rho(ds\times d\tau)\mu_0(ds').$$

Thus we can define the measures $\rho(s', d\tau)$ by

$$\int_{\Omega_{\Lambda^c}} f(s'\times\tau)\rho(s', d\tau) = \int_{\Omega} e^{W(s,s'\times\tau)} f(s'\times\tau)\rho(ds\times d\tau)$$

i.e.,

$$\rho(s',h) = \int_{\Omega} h(\tau) e^{W(s,s'\times\tau)}\rho(ds\times d\tau) = \rho(h e^{-W(s',\cdot)})$$

for $h \in C(\Omega_{\Lambda^c})$, where we have used $W(s,s'\times\tau) = -W(s',s\times\tau)$. Now since $W(s_1, s'\times\tau) = W(s_2, s'\times\tau) + W(s_1, s_2\times\tau)$ we have

$$\rho(s_1, h) = \rho(h e^{-W(s_1,\cdot)}) = \rho(s_2, h e^{-W(s_1, s_2\times\cdot)})$$

i.e.,

$$\rho(s_1, d\tau) = e^{-W(s_1, s_2\times\tau)}\rho(s_2, d\tau).$$

Thus the equations (6) imply the DLR equations (3).

Conversely, suppose ρ satisfies the DLR equations. Then for any $s \in \Omega_\Lambda$ and $f \in C(\Omega)$

$$\rho(f) = \int_{\Omega_\Lambda}\int_{\Omega_{\Lambda^c}} f(s'\times\tau)\rho(s',d\tau)\mu_0(ds')$$

$$= \int_{\Omega_\Lambda}\int_{\Omega_{\Lambda^c}} e^{W(s,s'\times\tau)} f(s'\times\tau)\rho(s,d\tau)\mu_0(ds')\ .$$

By averaging over s we obtain

$$\rho(f) = \int_{\Omega_\Lambda}\int_{\Omega_\Lambda}\int_{\Omega_{\Lambda^c}} e^{W(s,s'\times\tau)} f(s'\times\tau)\rho(s,d\tau)\mu_0(ds')\mu_0(ds)$$

$$= \int_{\Omega_\Lambda}\int_{\Omega_{\Lambda^c}} T_\Lambda f(s\times\tau)\rho(s,d\tau)\mu_0(ds) = \rho(T_\Lambda f)$$

so the equations (6) are equivalent to the DLR equations as claimed, and we can take $F = f - T_\Lambda f$. It will be convenient when working in a torus T_n to have $f \in C(\Omega_{T_n})$, so we will assume $f \in C(\Omega_{\Lambda'})$ for some finite $\Lambda' \subset Z^\nu$ (with T_n being identified with a cube C_n containing Λ'). By continuity it is enough to prove $\rho(f - T_\Lambda f) = 0$ for such functions. Again we take $\tilde{F} \in C(\Omega_{\tilde{\Lambda}})$ with $\tilde{\Lambda} \supset \Lambda' \cup \Lambda$ finite and $\|F - \tilde{F}\|_\infty \le \varepsilon$. Proceeding with our method, for the torus T_n (identified with a cube C_n containing $\tilde{\Lambda}$) we can take $F_{n,\lambda} = f - T_\lambda f \in C(\Omega_{T_n})$ with

$$T_\lambda f(s\times\tau) = \int_{\Omega_\Lambda} e^{W_\lambda(s,s'\times\tau)} f(s'\times\tau)\mu_0(ds')\ s \in \Omega_\Lambda,\ \tau \in \Omega_{T_n}\setminus\Lambda$$

$$\begin{aligned} W_\lambda(s,s'\times\tau) &= H^{\Phi+\lambda\Psi}_{T_n}(s\times\tau) - H^{\Phi+\lambda\Psi}_{T_n}(s'\times\tau) \\ &= H^{\Phi}_{T_n}(s\times\tau) - H^{\Phi}_{T_n}(s'\times\tau) + \lambda \sum_{(i+\tilde{\Lambda})\cap\Lambda\neq\emptyset} [\tau_i\tilde{F}(s\times\tau) - \tau_i\tilde{F}(s'\times\tau)]\ . \end{aligned}$$

It is then easy to compute $\langle F_{n,\lambda} \rangle_{n,\lambda} = 0$. Just as in the discrete case we can make $\|W_\lambda - W\|_\infty$ arbitrarily small for $|\lambda|$ sufficiently small and n sufficiently large, and then

$$\|F_{n,\lambda} - F\|_\infty \le \|f\|_\infty \, e^{\|W\|_\infty} (e^{\|W_\lambda - W\|_\infty} - 1) \le \varepsilon .$$

This completes the proof. We state the result as

THEOREM III.2.1. *In the classical system, any invariant equilibrium state for an interaction* $\Phi \in \tilde{\mathcal{B}}$ *satisfies the DLR equations.*

III.3. *Time evolution and the KMS conditions*

Consider the quantum system in a finite set Λ, with Hamiltonian H_Λ. According to quantum mechanics, the time evolution of observables $A \in \mathfrak{A}_\Lambda$ is given by the *-automorphisms of $\mathfrak{A}_\Lambda$

$$\alpha_t^\Lambda(A) = e^{itH_\Lambda} A \, e^{-itH_\Lambda} \tag{8}$$

(*in units where* $\hbar = 1$). *Now the* $\mathfrak{A}_\Lambda$*-valued function* $\alpha_t(A)$ *extends to* an entire function $\alpha_z^\Lambda(A) = e^{izH_\Lambda} A \, e^{-izH_\Lambda}$. The Gibbs equilibrium state $\rho_1(A) = (\operatorname{tr} e^{-H_\Lambda})^{-1} \operatorname{tr}(e^{-H_\Lambda} A)$ then satisfies

$$\begin{aligned} \rho_1(A\alpha_t^\Lambda(B)) &= (\operatorname{tr} e^{-H_\Lambda})^{-1} \operatorname{tr}(e^{-H_\Lambda} A \, e^{itH_\Lambda} B \, e^{-itH_\Lambda}) \\ &= (\operatorname{tr} e^{-H_\Lambda})^{-1} \operatorname{tr}(e^{-H_\Lambda} e^{H_\Lambda} e^{itH_\Lambda} B \, e^{-itH_\Lambda} e^{-H_\Lambda} A) \\ &= \rho_1(\alpha_{t-i}^\Lambda(B) A) \qquad \text{for any } A, B \in \mathfrak{A}_\Lambda . \end{aligned}$$

Conversely, suppose ρ is a state on $\mathfrak{A}_\Lambda$ with

$$\rho(A\alpha_t^\Lambda(B)) = \rho(\alpha_{t-i}^\Lambda(B) A) \qquad \text{for all } A, B \in \mathfrak{A}_\Lambda . \tag{9}$$

There must be a density matrix $\rho^{(\Lambda)}$ for ρ, i.e., a self-adjoint member of $\mathfrak{A}_\Lambda$ with $\operatorname{tr}(\rho^{(\Lambda)} A) = \rho(A)$ for all $A \in \mathfrak{A}_\Lambda$. Taking $t = 0$ in (9) we have

$$\mathrm{tr}(\rho^{(\Lambda)}AB) = \mathrm{tr}(\rho^{(\Lambda)}e^{H_\Lambda}B\,e^{-H_\Lambda}A) = \mathrm{tr}(e^{-H_\Lambda}A\,\rho^{(\Lambda)}e^{H_\Lambda}B)\,.$$

Since this is true for all B, we must have

$$e^{-H_\Lambda}A\,\rho^{(\Lambda)}e^{H_\Lambda} = \rho^{(\Lambda)}A \quad \text{for all } A \in \mathfrak{A}_\Lambda\,. \tag{10}$$

Taking A as the identity matrix 1, we find that $\rho^{(\Lambda)}$ commutes with e^{H_Λ}. Then (10) says that $\rho^{(\Lambda)}e^{H_\Lambda}$ commutes with every member of $\mathfrak{A}_\Lambda$, and so must be proportional to 1. Thus $\rho^{(\Lambda)} = (\mathrm{tr}\,e^{-H_\Lambda})^{-1}e^{-H_\Lambda}$ and $\rho = \rho_1$. So for the finite system, the equations (9) uniquely characterize the equilibrium state.

More generally, we can consider any C^*-algebra $\mathfrak{A}$ with a strongly continuous 1-parameter group of automorphisms α_t. The functions $t \mapsto \alpha_t(A)$ may not all extend to entire $\mathfrak{A}$-valued functions, but there will be a dense subset $\tilde{\mathfrak{A}}$ of $\mathfrak{A}$ consisting of those A for which such an extension is possible. Specifically, suppose f is the holomorphic Fourier transform of a smooth function $\check{f}$ of compact support on $\mathbf{R}$. For each $A \in \mathfrak{A}$ let $A_f = \int_{\mathbf{R}} f(t)\alpha_t(A)\,dt$. Then the entire function $\alpha_z(A_f)$ is given by

$$\alpha_z(A_f) = \int_{\mathbf{R}} f(t-z)\alpha_t(A)\,dt$$

(note that this agrees with $\alpha_s(A_f)$ when $z = s \in \mathbf{R}$). Thus $A_f \in \tilde{\mathfrak{A}}$. Since Fourier transforms of smooth functions of compact support are dense in the Schwartz space $\mathcal{S}(\mathbf{R})$, we can approximate A by elements A_f of $\tilde{\mathfrak{A}}$, showing that $\tilde{\mathfrak{A}}$ is dense in $\mathfrak{A}$. We will say that the state ρ of $\mathfrak{A}$ satisfies the *KMS conditions* for the group of automorphisms α_t if

$$\rho(AB) = \rho(\alpha_{-i}(B)A) \quad \text{for all } A \in \mathfrak{A},\ B \in \tilde{\mathfrak{A}}\,. \tag{11}$$

This is one of several equivalent definitions (see [14]). Note that by substituting $\alpha_t(B)$ for B in (11) we obtain $\rho(A\alpha_t(B)) = \rho(\alpha_{t-i}(B)A)$ as in (9).

Another useful form of the definition is the following: ρ satisfies the KMS conditions for α_t if for each $A, B \in \mathfrak{A}$ there is a function $F_{AB}(z)$, continuous and bounded on the strip $0 \leq \operatorname{Im} z \leq 1$ and analytic in its interior, such that

(12) $\rho(A\alpha_t(B)) = F_{AB}(t)$ and $\rho(\alpha_t(B)A) = F_{AB}(t+i)$ for $t \in \mathbf{R}$.

This form of the KMS conditions is particularly useful because it will provide a setting for the application of the Phragmén-Lindelöf principle and related results. Here is a third definition: ρ satisfies the KMS conditions for α_t if

$$(13) \qquad \int_{\mathbf{R}} f(t-i)\,\rho(A\alpha_t(B))\,dt = \int_{\mathbf{R}} f(t)\,\rho(\alpha_t(B)A)\,dt$$

for all $A, B \in \mathfrak{A}$ and f the holomorphic Fourier transform of a smooth function $\check{f}$ of compact support.

LEMMA III.3.1. *The three definitions above are equivalent.*

Proof. Suppose ρ satisfies the first definition. Equation (13) simply says $\rho(A\alpha_i(B_f)) = \rho(B_f A)$, which is true by (11) since $B_f \in \tilde{\mathfrak{A}}$. So the first definition implies the third. If ρ satisfies the third definition, we can take $F_{AB_f}(z) = \rho(A\alpha_z(B_f))$, and (12) is satisfied. For an arbitrary $B \in \mathfrak{A}$, we can take a sequence $B_f \to B$. Note that α_t is an isometry of $\mathfrak{A}$, so that $\rho(A\alpha_t(B_f))$ and $\rho(\alpha_t(B_f)A)$ converge uniformly in t to $\rho(A\alpha_t(B))$ and $\rho(\alpha_t(B)A)$ respectively. By the Phragmén-Lindelöf principle the functions $F_{AB_f}(z)$ converge uniformly on the whole strip $0 \leq \operatorname{Im} z \leq 1$, so their limit $F_{AB}(z)$ is continuous, bounded and analytic in the interior. Thus the third definition implies the second. Finally, suppose ρ satisfies the second definition. Then if $B \in \tilde{\mathfrak{A}}$ and $A \in \mathfrak{A}$, the function $\rho(\alpha_z(B)A)$ is entire. Since $F_{AB}(z+i) = \rho(\alpha_z(B)A)$ for real z, the same equation must hold for all z in the strip. Taking $z = -i$ yields (11). ∎

LEMMA III.3.2. *If ρ satisfies the KMS conditions for α_t and $\mathfrak{A}$ has an identity, then ρ is invariant under α_t.*

Proof. Take $A = 1$ in (12). Then $F_{1B}(t+i) = \rho(\alpha_t(B)) = F_{1B}(t)$, so $F_{1B}(z)$ extends to a periodic function on the whole complex plane. This function is entire by Morera's Theorem (its integral around any closed path is zero). But by Liouville's Theorem any bounded entire function is constant. ∎

A state ρ on a C^*-algebra $\mathfrak{A}$ is said to be *faithful* if $A \in \mathfrak{A}$ and $\rho(A^*A) = 0$ imply $A = 0$. A DLR state of the classical lattice system is easily seen to be faithful on $C(\Omega)$, because for each finite Λ we have $\rho^{(\Lambda)} \geq e^{-2\|\Phi\|_{\sim}}$. The analogous result for the quantum system uses the fact that the quasilocal algebra $\mathfrak{A}$ is *simple*, i.e., has no nontrivial two-sided ideals.

LEMMA III.3.3 (see [31], Theorem 21, and [32]). *The quasilocal algebra $\mathfrak{A}$ of the quantum lattice system is simple. Any state of $\mathfrak{A}$ (or any simple C^*-algebra) satisfying the KMS conditions for some 1-parameter group α_t is faithful.*

Proof. It is clear that the local algebras $\mathfrak{A}_\Lambda$, consisting of all linear operators on the finite-dimensional space $\mathcal{H}_\Lambda$, are simple. Thus if J is a proper two-sided ideal of $\mathfrak{A}$ we have $\mathfrak{A}_\Lambda \cap J = \{0\}$. Now $\mathfrak{A}_\Lambda$ is still a Banach algebra under the new norm $A \mapsto \|A+J\|$. If $A \in \mathfrak{A}_\Lambda$ is self-adjoint, $\|A\|$ is equal to the spectral radius of A, which is at most $\|A+J\|$. By continuity we have $\|A\| \leq \|A+J\|$ for any self-adjoint $A \in \mathfrak{A}$. If $B \in J$, $B^*B \in J$ is self-adjoint, so $B^*B = 0$, and thus $B = 0$. Therefore $\mathfrak{A}$ is simple.

Now let ρ satisfy the KMS conditions for α_t. Consider the inner product $\langle C, D\rangle = \rho(C^*D)$ on A. If $\rho(A^*A) = 0$, the Cauchy-Schwartz Inequality shows that $\rho(BA) = 0$ for all $B \in \mathfrak{A}$. Thus

$$J = \{A \in \mathfrak{A} : \rho(A^*A) = 0\} = \{A \in \mathfrak{A} : \rho(BA) = 0 \text{ for all } B \in \mathfrak{A}\}$$

is a left ideal in $\mathfrak{A}$. But if $A \in J$ and $B \in \tilde{\mathfrak{A}}$,

$$\rho((AB)^*AB) = \rho(\alpha_{-i}(B)(AB)^*A) = 0 .$$

By continuity, $\rho((AB)^*AB) = 0$ for all $B \in \mathfrak{A}$, so J is a two-sided ideal. Since $\mathfrak{A}$ is simple, $J = \{0\}$. ∎

The Tomita-Takesaki theory (see [30]) shows that any faithful state satisfies the KMS conditions for some strongly continuous one-parameter group of *-automorphisms, not necessarily of $\mathfrak{A}$, but of the von Neumann algebra $\pi(\mathfrak{A})''$ in the GNS representation (which will be defined in Lemma IV.2.1).

THEOREM III.3.4 (see [30], Theorem 6.2). *If a faithful state* ρ *satisfies the KMS conditions for two one-parameter groups* σ_t *and* α_t *on* $\mathfrak{A}$, *(and* $\mathfrak{A}$ *has an identity), then* $\sigma_t = \alpha_t$.

Proof. Let $A, B \in \mathfrak{A}$ such that $\sigma_t(A)$ and $\alpha_t(B)$ extend to entire functions $\sigma_z(A)$ and $\alpha_z(B)$. Let $F(z,w) = \rho(\sigma_z(A)\alpha_w(B))$. For t real

$$F(t+i,w) = \rho(\sigma_{t+i}(A)\alpha_w(B)) = \rho(\alpha_w(B)\sigma_t(A))$$

by the KMS conditions for σ_t. Now by the KMS conditions for α_t

$$F(t+i, s+i) = \rho(\alpha_{s+i}(B)\sigma_t(A)) = \rho(\sigma_t(A)\alpha_s(B)) \text{ for } s,t \text{ real} .$$

Thus $F(z) \equiv F(z,z)$, which is bounded on the strip $0 \le \text{Im } z \le 1$ by $\sup\{\|\sigma_{it}(A)\| \, \|\alpha_{it}(B)\| : 0 \le t \le 1\}$, satisfies the periodicity condition $F(t+i) = F(t)$ and so is bounded on the whole plane. By Liouville's Theorem $F(z)$ is constant, so

$$\rho(A\sigma_{-t} \circ \alpha_t(B)) = \rho(\sigma_t(A)\alpha_t(B)) = \rho(AB) .$$

Since the sets of A and B as above are dense in $\mathfrak{A}$, this is true for any $A, B \in \mathfrak{A}$ by continuity. Since ρ is faithful we obtain $\sigma_{-t} \circ \alpha_t(B) = B$ for all $B \in \mathfrak{A}$. ■

We have not yet defined the time evolution of the infinite system for which we will use the KMS conditions. Clearly we should consider a limit of the finite-system evolution given by (8). In order to obtain such a limit, a more restrictive Banach space of interactions must be considered. We let $\mathcal{B}_r$ for $r > 0$ be the Banach space of interactions Φ of the quantum lattice system with norm

$$\|\Phi\|_r = \sum_{X \ni 0} e^{r|X|} \|\Phi(X)\| < \infty . \tag{14}$$

Note that the time evolution of the finite system can be expanded in terms of commutators as

$$e^{itH_\Lambda} A e^{-itH_\Lambda} = \sum_{n=0}^{\infty} \frac{(it)^n}{n!} [H_\Lambda, A]^{(n)} \tag{15}$$

where

$$[B,A]^{(0)} = A, \quad [B,A]^{(1)} = [B,A] = BA - AB ,$$

and

$$[B,A]^{(n+1)} = [B, [B,A]^{(n)}] .$$

This formula is easily verified by taking the derivative of each side with respect to t. We now prove an estimate to use on the right-hand side of (15).

LEMMA III.3.5 (see [25], Lemma 7.6.1). *If* $\Phi \in \mathcal{B}_r$ *and* $\Lambda' \subset \Lambda$ *finite with* $A \in \mathfrak{A}_{\Lambda'}$, *then*

$$\|[H_\Lambda^\Phi, A]^{(n)}\| \leq \|A\| e^{r|\Lambda'|} n! \, (2r^{-1} e^{-r} \|\Phi\|_r)^n . \tag{16}$$

Proof. We write out (using $H_\Lambda^\Phi = \sum_{X \subset \Lambda} \Phi(X)$)

$$[H_\Lambda^\Phi, A]^{(n)} = \sum_{X_1 \subset \Lambda} \cdots \sum_{X_n \subset \Lambda} [\Phi(X_n), [\cdots [\Phi(X_1), A] \cdots]] .$$

Now since local algebras for disjoint subsets of Z^ν commute with each other, the sum can be restricted to those X_i with $X_i \cap (\Lambda' \cup \bigcup_{j=1}^{i-1} X_j) \neq \emptyset$. For fixed $X_1, \cdots, X_{i-1}$ the number of translates of a given X which contribute as X_i is at most $|X| [|\Lambda'| + \sum_{j=1}^{i-1} (|X_j| - 1)]$; here $|X_j| - 1$ appears rather than $|X_j|$ because each X_j must share at least one site with $\Lambda' \cup X_1 \cup \cdots \cup X_{j-1}$ in order for a nonzero term to occur. Thus we have

$$\|[H_\Lambda^\Phi, A]^{(n)}\| \leq \sum_{X_1 \ni 0} \cdots \sum_{X_n \ni 0} 2^n \|A\| \prod_{i=1}^{n} \left(\left[|\Lambda'| + \sum_{j=1}^{i-1} (|X_j| - 1) \right] \|\Phi(X_i)\| \right)$$

$$= 2^n \|A\| \sum_{k_1=0}^{\infty} \cdots \sum_{k_n=0}^{\infty} \prod_{i=1}^{n} \left[\left(|\Lambda'| + \sum_{j=1}^{i-1} k_j \right) \sum_{\substack{X_i \ni 0 \\ |X_i| = k_i + 1}} \|\Phi(X_i)\| \right] .$$

Now

$$\prod_{i=1}^{n} \left(|\Lambda'| + \sum_{j=1}^{i=1} k_j \right) \leq \left(|\Lambda'| + \sum_{j=1}^{n} k_j \right)^n \leq n!\, r^{-n} \exp\left(r|\Lambda'| + r \sum_{j=1}^{n} k_j \right)$$

$$= n!\, r^{-n} e^{r|\Lambda'|} \prod_{j=1}^{n} \exp(rk_j)$$

so

$$\|[H_\Lambda^\Phi, A]^{(n)}\| \le 2^n \|A\|\, n!\, r^{-n} e^{r|\Lambda'|} \sum_{k_1=0}^{\infty} \cdots \sum_{k_n=0}^{\infty} \prod_{j=1}^{n} \left[\exp(rk_j) \times \sum_{\substack{X_j \ni 0 \\ |X_j| = k_j+1}} \|\Phi(X_j)\| \right]$$

$$= 2^n \|A\|\, n!\, r^{-n} e^{r|\Lambda'|} \left(\sum_{X \ni 0} e^{r(|X|-1)} \|\Phi(X)\| \right)^n$$

$$= \|A\|\, n!\, e^{r|\Lambda'|} (2 r^{-1} e^{-r} \|\Phi\|_r)^n .$$ ∎

The point to notice about this estimate is that it is independent of Λ. Thus as $\Lambda \to \infty$ (eventually containing each lattice site) we have $[H_\Lambda^\Phi, A]^{(n)} \to \sum_{X_1 \subset Z^\nu} \cdots \sum_{X_n \subset Z^\nu} [\Phi(X_n), [\cdots [\Phi(X_1), A] \cdots]]$ (this sum being absolutely convergent). Now instead of the "free boundary condition" Hamiltonian H_Λ^Φ we could consider external-field or periodic boundary conditions as in equations (25) and (26) of Section I.3. A similar argument to that above shows that (16) is also valid for the Hamiltonians ${}_hH_\Lambda^\Phi$ and $H_{T_m}^\Phi$ (with $\Lambda' \subset T_m$), and further that for any sequence of external-field boundary conditions h with $\Lambda \to \infty$, or for a sequence of tori T_m (identified with cubes $C_m \to \infty$), the same limit $\sum_{X_1 \subset Z^\nu} \cdots \sum_{X_n \subset Z^\nu} [\Phi(X_n), [\cdots [\Phi(X_1), A] \cdots]]$ is obtained.

THEOREM III.3.6 (see [25], Theorem 7.6.2). *If $\Phi \in \mathcal{B}_r$ for some $r > 0$, there is a strongly continuous one-parameter group α_t^Φ of *-automorphisms of $\mathfrak{A}$ given as follows: if $A \in \mathfrak{A}_{\Lambda'}$, for some finite $\Lambda' \subset Z^\nu$ and $\Lambda \to \infty$ (every site being eventually contained in Λ), then*

$$\lim_{\Lambda \to \infty} \exp(itH_\Lambda^\Phi) A \exp(-itH_\Lambda^\Phi) = \alpha_t^\Phi(A) . \tag{17}$$

The same limit is obtained by replacing H_Λ^Φ *in* (17) *by* ${}_hH_\Lambda^\Phi$ *for any external-field boundary conditions* h, *or by* $H_{T_m}^\Phi$ *for tori* $T_m = (Z/mZ)^\nu$ *identified with cubes* C_m *tending to infinity in the same way. The convergence is uniform in* t *on any bounded interval. If* $A \in \mathfrak{A}_{\Lambda'}$ (Λ' *finite*) *and* $|t| < r\, e^r (2\|\Phi\|_r)^{-1}$ *we have*

$$(18)\quad \alpha_t^\Phi(A) = \sum_{n=0}^{\infty} (n!)^{-1} (it)^n \sum_{X_1 \subset Z^\nu} \cdots \sum_{X_n \subset Z^\nu} [\Phi(X_n),[\cdots[\Phi(X_1), A]\cdots]] \,.$$

Proof. By the estimate (16) the series (18) converges absolutely for $|t| < T = r\, e^r (2\|\Phi\|_r)^{-1}$, and is the limit of the series (15) (using either H_Λ^Φ, ${}_hH_\Lambda^\Phi$ or $H_{T_m}^\Phi$ as Hamiltonian) as $\Lambda \to \infty$ (or in the periodic case $C_m \to \infty$). The convergence is uniform in t on any closed subinterval of $(-T,T)$. Since α_t^Φ (being given by (17)) is a *-isomorphism (and hence an isometry) on each $\mathfrak{A}_{\Lambda'}$, it extends by continuity to a *-automorphism of $\mathfrak{A}$. It is clear that $\lim_{t \to 0} \alpha_t^\Phi(A) = A$ (first for $A \in \mathfrak{A}_{\Lambda'}$ and then by a "3ε" argument for all A), and that $\alpha_t \circ \alpha_s(A) = \alpha_{t+s}(A)$ when $|t|$, $|s|$ and $|t+s|$ are all less than T. We now define $\alpha_t^\Phi(A) = (\alpha_{t/k}^\Phi)^k(A)$ for all $t \in \mathbf{R}$, taking $kT > |t|$. All the desired properties of α_t^Φ are easily seen to be preserved by this definition; in particular, (17) is true for all $t \in \mathbf{R}$, the convergence being uniform on any bounded interval. ∎

COROLLARY III.3.7. *For any* $A \in \mathfrak{A}$ *the maps* $\Phi \mapsto \alpha_t^\Phi(A)$ *are continuous from* $\mathcal{B}_r$ *to* $\mathfrak{A}$, *and equicontinuous for* t *in any bounded interval.*

Proof. For $A \in \bigcup_{|\Lambda| < \infty} \mathfrak{A}_\Lambda$ and t in a closed subinterval of $(-T,T)$ this is clear, since $\alpha_t^\Phi(A)$ is given by the absolutely convergent power series (18). The usual "3ε" argument shows that this also holds for any $A \in \mathfrak{A}$. We can extend the result to allow any bounded interval for t, using

$$\|\alpha^{\Psi}_{t+s}(A)-\alpha^{\Phi}_{t+s}(A)\| \le \|\alpha^{\Psi}_{t}[\alpha^{\Psi}_{s}(A)-\alpha^{\Phi}_{s}(A)]\| + \|\alpha^{\Psi}_{t}(\alpha^{\Phi}_{s}(A))-\alpha^{\Phi}_{t}(\alpha^{\Phi}_{s}(A))\|$$

and the fact that α^{Ψ}_{t} is an isometry. ■

A state ρ of $\mathfrak{A}$ will be called a *KMS state* for $\Phi \in \mathfrak{B}_r$ if it satisfies the KMS conditions for the time evolution α^{Φ}_{t}. We now apply the methods of Section III.2 to show that invariant equilibrium states for interactions in $\mathfrak{B}_r$ are KMS states. This was proven in [17] using indirect methods. Araki [2] has recently proven the converse: invariant KMS states are invariant equilibrium states.

THEOREM III.3.8. *If* $\Phi \in \mathfrak{B}_r$ *for some* $r > 0$, *every invariant equilibrium state for* Φ *is a KMS state for* Φ.

Proof. We will use the definition of KMS conditions by equation (13). By continuity it will be enough to prove this for $A, B \in \mathfrak{A}_{\Lambda'}$ for some finite Λ'. Then (13) is almost in the desired form $\rho(F) = 0$, except that we need F to be self-adjoint: an interaction must have $\Psi(X)$ self-adjoint. Therefore we take real and imaginary parts, letting

$$F = \mathrm{Re} \int_{\mathbf{R}} [f(t-i)\, A\alpha^{\Phi}_{t}(B) - f(t)\alpha^{\Phi}_{t}(B)\, A]\, dt \tag{19}$$

where $A, B \in \mathfrak{A}_{\Lambda'}$ and f is the holomorphic Fourier transform of a smooth function $\check{f}$ of compact support on $\mathbf{R}$. If $\rho(F) = 0$ for all such F, then ρ will be a KMS state for Φ.

The obvious choice for $F_{n,\lambda}$ (with n large enough that $\tilde{\Lambda} \subset C_n$ as in Section III.2) is

$$F_{n,\lambda} = \mathrm{Re} \int_{\mathbf{R}} [f(t-i)\, A\alpha^{n,\lambda}_{t}(B) - f(t)\alpha^{n,\lambda}_{t}(B)\, A]\, dt$$

where

$$\alpha^{n,\lambda}_{t}(B) = \exp(itH^{\Phi+\lambda\Psi}_{T_n})\, B \exp(-itH^{\Phi+\lambda\Psi}_{T_n}) .$$

Since the Gibbs equilibrium state $\langle\cdot\rangle_{n,\lambda}$ for $\Phi+\lambda\Psi$ in T_n satisfies the KMS conditions for the time evolution $\alpha_t^{n,\lambda}$ we have $\langle F_{n,\lambda}\rangle_{n,\lambda} = 0$.

To complete the proof, we must show that given $\varepsilon > 0$ there are $\eta > 0$ and N such that for $|\lambda| < \eta$ and $n > N$ we have $\|F - F_{n,\lambda}\| < \varepsilon$. Since the restrictions of f to the lines $\text{Im } z = 0$ and $\text{Im } z = -1$ are in L^1 it is enough to prove that for any M and $\delta > 0$ there are $\eta > 0$ and N such that $\|\alpha_t^{\Phi}(B) - \alpha_t^{n,\lambda}(B)\| < \delta$ for $|t| \le M$, $|\lambda| < \eta$ and $n > N$. But the convergence of $\alpha_t^{n,\lambda}(B)$ to $\alpha_t^{\Phi+\lambda\Psi}(B)$ as $n \to \infty$ is easily seen to be uniform in both t and λ on bounded sets, while by Corollary III.3.7 the maps $\lambda \mapsto \alpha_t^{\Phi+\lambda\Psi}(B)$ are equicontinuous in $t \in [-M,M]$. This completes the proof. ∎

III.4. *Physical equivalence and strict convexity*

As we have seen in Section I.4, if two interactions $\Phi, \Psi \in \mathcal{B}$ are physically equivalent the pressure P is linear on the line through Φ and Ψ, and so Φ and Ψ have the same invariant equilibrium states. In this section physical equivalence is examined from the point of view of the DLR and KMS conditions. We find that physically equivalent interactions in the appropriate Banach spaces have (in the classical case) the same DLR states, or (in the quantum case) the same time evolution and KMS states, providing further evidence that "physical equivalence" has been correctly defined. Conversely, we show that DLR and KMS states uniquely determine the interaction up to physical equivalence. This will show that the pressure P is strictly convex on $\tilde{\mathcal{B}}$ (classical case) or $\mathcal{B}_r$ (quantum case) in directions not corresponding to physical equivalence. The first results in this direction are due to Griffiths and Ruelle [12], who showed that in the space $\tilde{\mathcal{B}}^G$ of classical "lattice-gas" interactions and the space $\mathcal{B}_r^S$ of quantum interactions with $\text{tr}_Y(\Phi(X)) = 0$ for all $X \supset Y \neq \emptyset$, a state can satisfy the DLR equations or KMS conditions respectively for at most one interaction. The quantum case was further developed by Roos [23], whose result is essentially Theorem III.4.2 below.

THEOREM III.4.1. *In the classical lattice system, if* Φ *and* Ψ *are in* $\tilde{\mathcal{B}}$ *the following are equivalent:*

(i) Φ *and* Ψ *are physically equivalent, i.e.,* $S(\Phi-\Psi) = 0$

(ii) P *is linear on the line segment between* Φ *and* Ψ

(iii) *Some state* ρ *satisfies the DLR equations for both* Φ *and* Ψ

(iv) *Every state satisfying the DLR equations for* Φ *does so for* Ψ, *and vice versa.*

Proof. We will have the following implications:

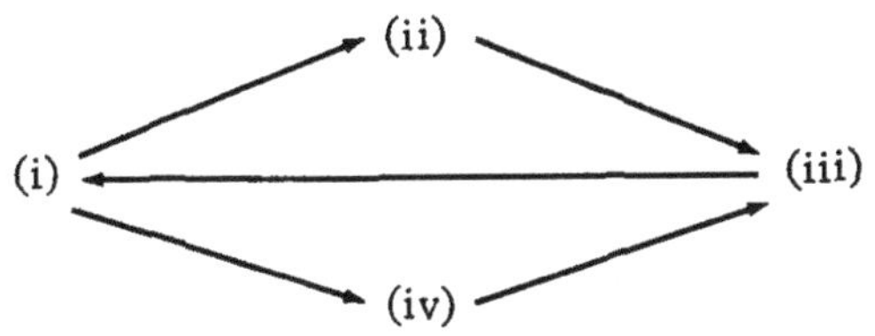

(i) → (ii) is true by Theorem I.4.1, while (iv) → (iii) is trivial. If (ii) holds, by the Hahn-Banach Theorem there is a linear functional on $\tilde{\mathcal{B}}$ tangent to P at both Φ and Ψ (and on the segment between), and by Theorem III.2.1 the invariant equilibrium state it determines satisfies the DLR equations for both Φ and Ψ. Thus (ii) → (iii).

To show (iii) → (i) → (iv), we consider the functions $W^{\Phi}(s,\omega)$ and $W^{\Psi}(s,\omega)$ as in Section III.1, where

$$W^{\Phi}(s_1, s_2\times\tau) = \sum_{X\cap\Lambda\neq\emptyset} [\Phi(X)(s_1\times\tau) - \Phi(X)(s_2\times\tau)]$$

for some finite $\Lambda \subset \mathbf{Z}^{\nu}$, $s_1, s_2 \in \Omega_{\Lambda}$ and $\tau \in \Omega_{\Lambda^c}$. It is clear from the definition of the DLR equations (3) that if $W^{\Phi} = W^{\Psi}$ for every Λ, Φ and Ψ will have the same DLR states. On the other hand, the functions W^{Φ} are uniquely determined by the DLR state ρ, since $\exp(-W^{\Phi}(s_1,s_2\times\cdot))$ is the Radon-Nikodym derivative $\rho(s_1,d\tau)/\rho(s_2,d\tau)$ and is continuous (note that the measure $\rho(s_2,d\tau)$ cannot be zero on any nonempty open set in Ω_{Λ^c}). Thus it will suffice to show that $W^\Phi = W^\Psi$ if and only if Φ and Ψ are physically equivalent. To simplify matters further, it will suffice to consider $\Lambda = \{0\}$; all the functions W^Φ for other Λ are determined by that for $\Lambda = \{0\}$ since we can go from one configuration in Λ to any other in a finite number of steps by changing one site at a time. For $s_1, s_2 \in \Omega_0$ we will write $W^\Phi(s_1, s_2 \times \tau)$ as $W^\Phi_{12}(\tau)$ to simplify notation.

It will be instructive to treat the easiest case first, namely the Ising model, $\Omega_0 = \{-1, +1\}$, μ_0 normalized counting measure. We need only consider $s_1 = +1$, $s_2 = -1$. Treating Ω as a compact group with characters $\sigma_X(X \subset Z^\nu$ finite), the Fourier coefficients of W^Φ_{12} are

$$\hat{W}^\Phi_{12}(Y) = \langle W^\Phi_{12}\sigma_Y\rangle_0 = 2\sum_{X \ni 0} \langle \Phi(X)\sigma_{Y\cup\{0\}}\rangle_0$$

where $0 \not\subset Y$. Note that only terms with $Y \cup \{0\} \subset X$ contribute to the above sum. Thus by the definition of $S\Phi$ (equation (35) of Section I.4), we have

$$S\Phi(Y\cup\{0\}) = \sum_{X \supset Y\cup\{0\}} \langle \Phi(X)\sigma_{Y\cup\{0\}}\rangle_0\, \sigma_{Y\cup\{0\}} = \frac{1}{2}\hat{W}^\Phi_{12}(Y)\sigma_{Y\cup\{0\}}$$

so that $W^\Phi = W^\Psi$ if and only if $S(\Phi - \Psi) = 0$.

In the general case, for any $s_1, s_2 \in \Omega_0$ we obtain

$$E_Y W^\Phi_{12}(s') = \sum_{X \ni 0} ([E_{Y\cup\{0\}}\Phi(X)](s_1 \times s') - [E_{Y\cup\{0\}}\Phi(X)](s_2 \times s'))$$

for $Y \subset Z^\nu \backslash \{0\}$ finite and $s' \in \Omega_Y$. Note that $Q_Y\Phi(X)$ depends only on Ω_Y and is zero when $Y \not\subset X$. Thus for $Z \subset Z^\nu \backslash \{0\}$ finite we have

$$
\begin{aligned}
Q_Z W_{12}^{\Phi}(s') &= \sum_{Y \subset Z} (-1)^{|Z \setminus Y|} E_Y W_{12}^{\Phi}(s') \\
&= \sum_{X \supset Z \cup \{0\}} \left([Q_{Z \cup \{0\}} \Phi(X)](s_1 \times s') - [Q_{Z \cup \{0\}} \Phi(X)](s_2 \times s') \right) \\
&= S\Phi(Z \cup \{0\})(s_1 \times s') - S\Phi(Z \cup \{0\})(s_2 \times s') \qquad (\text{for } s' \in \Omega_Z)
\end{aligned}
$$

where $S\Phi$ is defined as in equation (40) of Section I.4. Thus if $S(\Phi - \Psi) = 0$ we have $W_{12}^{\Phi} = W_{12}^{\Psi}$ (since $W_{12}^{\Phi} = \sum_{Z \subset \mathbf{Z}^{\nu} \setminus \{0\}} Q_Z W_{12}^{\Phi}$). By integrating over s_2 we obtain

$$
S\Phi(Z \cup \{0\})(s_1 \times s') = \int_{\Omega_0} Q_Z W_{12}^{\Phi}(s') \mu_0(ds_2)
$$

so that the converse is also true. ■

THEOREM III.4.2 (see [23]). *In the quantum lattice system, if* Φ *and* Ψ *are in some* $\mathcal{B}_r$ *the following are equivalent:*

(i) Φ *and* Ψ *are physically equivalent*

(ii) P *is linear on the line segment between* Φ *and* Ψ

(iii) *Some state* ρ *is* KMS *for both* Φ *and* Ψ

(iv) *The time evolutions* α_t^{Φ} *and* α_t^{Ψ} *are equal*

(v) *Every* KMS *state for* Φ *is* KMS *for* Ψ, *and vice versa.*

Proof. We will have the following implications:

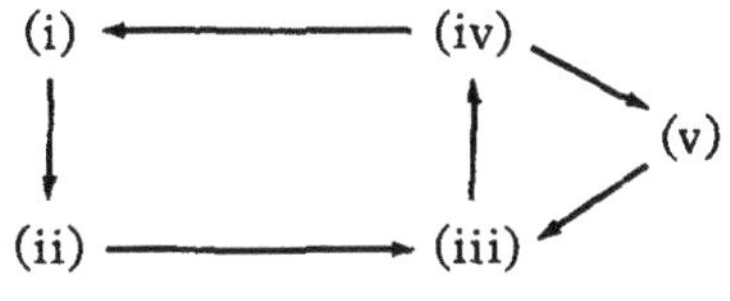

(i) → (ii) is true by Theorem I.4.2, (ii) → (iii) by the Hahn-Banach Theorem and Theorem III.3.8, and (iii) → (iv) by Theorem III.3.4, while (iv) → (v) → (iii) is trivial. It remains to be shown that (iv) implies (i). Our proof is essentially that in [12].

Suppose $\alpha_t^{\Phi} = \alpha_t^{\Psi}$. Then for any $A \in \mathfrak{A}_\Lambda$ with Λ finite, the power series (18) shows that

$$\sum_{X \subset Z^\nu} [\Phi(X) - \Psi(X), A] = -i \frac{d}{dt} (\alpha_t^{\Phi}(A) - \alpha_t^{\Psi}(A))_{|t=0} = 0 .$$

Thus it suffices to show that if $\sum_{X \subset Z^\nu} [\Phi(X), A] = 0$ for all $A \in \bigcup_{|\Lambda| < \infty} \mathfrak{A}_\Lambda$, then $S\Phi = 0$. Now by taking the partial trace

$$0 = \mathrm{tr}_{\Lambda^c} \sum_{X \subset Z^\nu} [\Phi(X), A] = \left[\sum_{X \cap \Lambda \neq \emptyset} \mathrm{tr}_{\Lambda^c} \Phi(X), A \right] \quad \text{for all } A \in \mathfrak{A}_\Lambda .$$

Since the algebra $\mathfrak{A}_\Lambda$ has trivial center, $\sum_{X \cap \Lambda \neq \emptyset} \mathrm{tr}_{\Lambda^c} \Phi(X)$ must be a scalar multiple of the identity. But then for any nonempty finite Y

$$S\Phi(Y) = Q_Y \sum_{X \supset Y} \Phi(X) = Q_Y \sum_{X \cap Y \neq \emptyset} \Phi(X)$$

$$= \sum_{Z \subset Y} (-1)^{|Y \setminus Z|} \mathrm{tr}_{Z^c} \sum_{X \cap Y \neq \emptyset} \Phi(X)$$

is again a scalar multiple of 1. Now this scalar must be zero, since for Y nonempty $\mathrm{tr} \circ Q_Y = 0$. ■

It is also possible to turn the above argument around, providing a direct proof that (i) implies (iv) (see [23], Lemma 3.4). First we show that if $S\Phi = 0$ and $A \in \bigcup_{|\Lambda| < \infty} \mathfrak{A}_\Lambda$ then $\sum_{X \subset Z^\nu} [\Phi(X), A] = 0$. For any finite Λ we have

$$\mathrm{tr}_{\Lambda^c} \sum_{X \cap \Lambda \neq \emptyset} \Phi(X) = \sum_{Y \subset \Lambda} Q_Y \sum_{X \cap \Lambda \neq \emptyset} \Phi(X) = \mathrm{tr} \sum_{X \cap \Lambda \neq \emptyset} \Phi(X) + \sum_{\emptyset \neq Y \subset \Lambda} S\Phi(Y)$$

so that if $S\Phi = 0$, $\mathrm{tr}_{\Lambda^c} \sum_{X \cap \Lambda \neq \emptyset} \Phi(X)$ is a scalar multiple of 1. Note that for any $B \in \mathfrak{A}$, $B = \lim_{\Lambda\to\infty} \operatorname{tr}_{\Lambda^c} B$ (this is trivial for $B \in \bigcup_{|\Lambda|<\infty} \mathfrak{A}_\Lambda$, and the "$3\varepsilon$" trick extends it to all of $\mathfrak{A}$). Then for any $A \in \bigcup_{|\Lambda|<\infty} \mathfrak{A}_\Lambda$

$$\sum_{X\subset Z^\nu} [\Phi(X),A] = \lim_{\Lambda\to\infty} \operatorname{tr}_{\Lambda^c}\left[\sum_{X\cap\Lambda\neq\emptyset} \Phi(X),A\right] = \lim_{\Lambda\to\infty}\left[\operatorname{tr}_{\Lambda^c} \sum_{X\cap\Lambda\neq\emptyset} \Phi(X),A\right] = 0 .$$

Now by (18) it is enough to show that for all $A \in \bigcup_{|\Lambda|<\infty} \mathfrak{A}_\Lambda$ and all n,

$$\sum_{X_1\subset Z^\nu} \cdots \sum_{X_n\subset Z^\nu} [\Phi(X_n), [\cdots[\Phi(X_1), A]\cdots]]$$

equals the corresponding expression for Ψ. The proof will be by induction on n; we have the $n=1$ case above. For the induction step, it suffices to show

$$\sum_{X_1\subset Z^\nu} \cdots \sum_{X_n\subset Z^\nu} [\Phi(X_n)-\Psi(X_n), [\Phi(X_{n-1}), [\cdots[\Phi(X_1), A]\cdots]]] = 0 .$$

Since the above sum converges absolutely by the estimates in Lemma III.3.5, it is the limit as $\Lambda\to\infty$ of

$$\sum_{X_n\subset Z^\nu} \sum_{X_1\subset\Lambda} \cdots \sum_{X_{n-1}\subset\Lambda} [\Phi(X_n)-\Psi(X_n), [\Phi(X_{n-1}), [\cdots[\Phi(X_1), A]\cdots]]]$$

which is zero as in the $n=1$ case, since $\sum_{X_1\subset\Lambda} \cdots \sum_{X_{n-1}\subset\Lambda} [\Phi(X_{n-1}), [\cdots[\Phi(X_1), A]\cdots]]$ is in $\mathfrak{A}_\Lambda$. This completes the proof.

III.5. *The KMS condition for classical interactions*

As in section I.3, we can consider a classical discrete-spin system (with μ_0 normalized counting measure) as a special case of a corresponding quantum system, involving diagonal matrices for a distinguished basis.

A state ρ on $C(\Omega)$ has a natural extension to a state ρ on $\mathfrak{A}$, whose restriction to $\mathfrak{A}_\Lambda$ is given by a diagonal matrix in the distinguished basis of $\mathcal{H}_\Lambda$. A state on $\mathfrak{A}$ will be called *classical* if it arises from a state on $C(\Omega)$ in this way. An equivalent formulation is obtained by taking matrix units E^Λ_{ij} in $\mathfrak{A}_\Lambda$ corresponding to the basis of $\mathcal{H}_\Lambda$ (thus $E^\Lambda_{ij} E^\Lambda_{k\ell} = \delta_{jk} E^\Lambda_{i\ell}$ and $(E^\Lambda_{ij})^* = E^\Lambda_{ji}$). Then the state ρ is classical if and only if for all finite Λ, $\rho(E^\Lambda_{ij}) = 0$ when $i \neq j$.

In this section we will show that for a classical interaction $\Phi \in \tilde{\mathcal{B}}$ (i.e., one with all $\Phi(X)$ diagonal in the distinguished basis) a state ρ on $\mathfrak{A}$ satisfies the KMS condition if and only if it is classical and satisfies the DLR equations. This result is due to Brascamp [36].

Of course, before we can use the KMS conditions we must know that our interaction Φ defines a suitable time evolution; in the general case this requires that Φ be in some $\mathcal{B}_r$, but we will now show that for a classical interaction, $\Phi \in \tilde{\mathcal{B}}$ is sufficient.

LEMMA III.5.1. *Let $\Phi \in \tilde{\mathcal{B}}$ be a classical interaction. Then there is a strongly continuous one-parameter group α^Φ_t of *-automorphisms of $\mathfrak{A}$ defined as follows: for $A \in \mathfrak{A}_\Lambda$ with Λ finite,*

$$\alpha^\Phi_t(A) = \exp(iK_\Lambda t)\, A \exp(-iK_\Lambda t) \tag{19}$$

where $K_\Lambda = \sum_{X \cap \Lambda \neq \emptyset} \Phi(X)$. If $\Phi \in \mathcal{B}_r$ this is the same time evolution that was defined in Theorem III.3.6.

Proof. Note that $K_\Lambda \in C(\Omega)$ with $\|K_\Lambda\| \leq |\Lambda|\, \|\Phi\|_{\sim}$. If $\Lambda \subset \Lambda'$, then $K_{\Lambda'} - K_\Lambda$ commutes with both K_Λ and $\mathfrak{A}_\Lambda$, so that $\exp(iK_{\Lambda'} t) \times A \exp(-iK_{\Lambda'} t) = \exp(iK_\Lambda t)\, A \exp(-iK_\Lambda t)$. Therefore (19) is a consistent definition of α^Φ_t on $\bigcup_{\Lambda \text{ finite}} \mathfrak{A}_\Lambda$. Since these are isometries there is a unique continuous extension to $\mathfrak{A}$. The usual "3ε" trick shows that α^Φ_t is a strongly continuous group of *-automorphisms of $\mathfrak{A}$. Note that $\alpha^\Phi_t(A)$

for $A \in \bigcup_{\Lambda \text{ finite}} \mathfrak{A}_\Lambda$ is also given by the power series (18), which is absolutely convergent in the whole complex plane. ∎

THEOREM III.5.2. *Let $\Phi \in \tilde{\mathfrak{B}}$ be a classical interaction. Then a state ρ on $\mathfrak{A}$ (not necessarily invariant) satisfies the KMS conditions for the time evolution α_t^Φ if and only if it is classical and satisfies the DLR equations.*

Proof. We write the DLR equations in the following form, which is simply a restatement of (3) in a language more suited to the quantum system: for any $\Lambda \subset \mathbf{Z}^\nu$ finite, $s_1, s_2 \in \Omega_\Lambda$, and $f \in C(\Omega_{\Lambda^c})$,

$$\rho(e^{K_\Lambda} f E_{11}^\Lambda) = \rho(e^{K_\Lambda} f E_{22}^\Lambda), \tag{20}$$

where E_{ii}^Λ is the projection in $\mathfrak{A}_\Lambda$ corresponding to s_i.

Suppose ρ is KMS. First we show that ρ is classical. Using the form (11) of the KMS conditions, and noting $\mathfrak{A}_\Lambda \subset \tilde{\mathfrak{A}}$ by (19), we have for $i \neq j$

$$\rho(E_{ij}^\Lambda) = \rho(E_{ij}^\Lambda E_{jj}^\Lambda) = \rho(\alpha_{-i}(E_{jj}^\Lambda) E_{ij}^\Lambda) = \rho(e^{K_\Lambda} E_{jj}^\Lambda e^{-K_\Lambda} E_{ij}^\Lambda) = \rho(E_{jj}^\Lambda E_{ij}^\Lambda) = 0$$

since the diagonal matrix E_{jj}^Λ commutes with K_Λ. Next, with Λ, E_{ii}^Λ and f as in (20) we have

$$\rho(e^{K_\Lambda} f E_{11}^\Lambda) = \rho(e^{K_\Lambda} f E_{12}^\Lambda E_{21}^\Lambda) = \rho(\alpha_{-i}(E_{21}^\Lambda) e^{K_\Lambda} f E_{12}^\Lambda)$$

$$= \rho(e^{K_\Lambda} E_{21}^\Lambda f E_{12}^\Lambda) = \rho(e^{K_\Lambda} f E_{22}^\Lambda)$$

because $f \in \mathfrak{A}_{\Lambda^c}$ and $E_{21}^\Lambda \in \mathfrak{A}_\Lambda$ commute. So ρ satisfies the DLR equations.

Conversely, suppose ρ is classical and satisfies (20). First we claim that $\rho(e^{K_\Lambda} B E_{11}^\Lambda) = \rho(e^{K_\Lambda} B E_{22}^\Lambda)$ for any $B \in \mathfrak{A}_{\Lambda^c}$. It is enough to prove this when B is a matrix unit $E^Y_{k\ell}$ for Y disjoint from Λ. If $k=\ell$ this is (20), while if $k \neq \ell$

$$\rho(e^{K_\Lambda} E^Y_{k\ell} E^\Lambda_{11}) = \rho(E^Y_{kk} E^\Lambda_{11} e^{K_\Lambda} E^Y_{k\ell} E^\Lambda_{11}) = 0$$

since $E^Y_{kk} E^\Lambda_{11} e^{K_\Lambda} E^Y_{k\ell} E^\Lambda_{11}$ has "off-diagonal matrix elements" only (to be precise, we approximate e^{K_Λ} by a diagonal matrix in some $\mathfrak{A}_X$, and look at matrix elements in $\mathfrak{A}_{\Lambda \cup X \cup Y}$).

We will prove the KMS conditions in the form (11) where A and B are matrix units E^Λ_{ij} and $E^\Lambda_{k\ell}$ in some $\mathfrak{A}_\Lambda$. The same sort of approximation argument as in Lemma III.3.1 will obtain equation (12) from this. Now $\rho(E^\Lambda_{ij} E^\Lambda_{k\ell}) = 0$ unless $j = k$ and $i = \ell$, in which case we have $\rho(E^\Lambda_{ii})$. On the other hand, unless $\ell = i$ and $k = j$,

$$\rho(\alpha_{-i}(E^\Lambda_{k\ell}) E^\Lambda_{ij}) = \rho(e^{K_\Lambda} E^\Lambda_{k\ell} e^{-K_\Lambda} E^\Lambda_{ij}) = \rho(E^\Lambda_{kk} e^{K_\Lambda} E^\Lambda_{k\ell} e^{-K_\Lambda} E^\Lambda_{\ell\ell} E^\Lambda_{ij}) = 0$$

(since for $k \neq j$ the matrix elements are off-diagonal). For $\ell = i$ and $k = j$ we obtain

$$\rho(e^{K_\Lambda} E^\Lambda_{ji} e^{-K_\Lambda} E^\Lambda_{ij}) = \sum_r \rho(e^{K_\Lambda} E^\Lambda_{ri} e^{-K_\Lambda} E^\Lambda_{ir} E^\Lambda_{jj}) .$$

Note that $\sum_r E^\Lambda_{ri} e^{-K_\Lambda} E^\Lambda_{ir}$ commutes with any matrix unit in $\mathfrak{A}_\Lambda$, so is in $\mathfrak{A}_{\Lambda^c}$. Therefore the above expression is equal to

$$\sum_r \rho(e^{K_\Lambda} E^\Lambda_{ri} e^{-K_\Lambda} E^\Lambda_{ir} E^\Lambda_{ii}) = \rho(e^{K_\Lambda} E^\Lambda_{ii} e^{-K_\Lambda} E^\Lambda_{ii}) = \rho(E^\Lambda_{ii})$$

as required. ∎

IV. DECOMPOSITION OF STATES

IV.1. *Ergodic states*

The invariant equilibrium states for an interaction $\Phi \in \mathcal{B}$ form a convex set Δ_Φ, compact in the weak-* topology. Thus it is natural from the mathematical point of view to consider the extreme points of Δ_Φ. Note that these are also extremal in the set E^I of all invariant states, since by the Variational Principle the invariant equilibrium states are those invariant states on which the affine function $\rho \mapsto s(\rho) - \rho(A_\Phi)$ attains its supremum. These extreme points turn out to be important from the physical point of view as well, because they represent "pure thermodynamic phases." The "pure phases" are characterized among invariant equilibrium states by the property that "macroscopic" quantities are given definite values. Thus an experimenter might take a "macroscopic" sample of material and measure the average of some microscopic observable over the sample (e.g., the magnetization of a crystal is the average of contributions from each lattice site). For a pure phase, each measurement should yield approximately the same value, the small statistical fluctuations (rarely detectable in practice) disappearing as the size of the sample increases. If on the contrary the measured values are spread out over a large range or show several peaks, the experimenter would conclude that he is dealing with a mixture of phases.

Mathematically, this is expressed as follows: given a self-adjoint observable A (in $\mathfrak{A}$ for the quantum system, or $C(\Omega)$ for the classical), we form the average

$$c_n(A) = n^{-\nu} \sum_{i \in C_n} \tau_i A \tag{1}$$

of translates of A over a cube C_n of side n. The invariant state ρ will be called *ergodic* if

(2) $$\lim_{n\to\infty} \rho(c_n(A)^2) = \rho(A)^2 \text{ for all self-adjoint } A \in \mathfrak{A}.$$

Note that the "variance" of measurements of $c_n(A)$ in the state ρ is $\rho((c_n(A)-\rho(A))^2) = \rho(c_n(A)^2) - \rho(A)^2$, so that ergodic equilibrium states represent pure phases. In the classical case this "variance" is the square of the norm of the function $c_n(A) - \rho(A) \in L^2(\Omega,\rho)$. Thus for an ergodic state the translation averages of continuous observables (and by continuity observables in $L^2(\Omega,\rho)$ also) tend in L^2 norm to the constant $\rho(A)$, the "phase-space" average. The stronger statement that the translation averages of L^1 observables converge almost everywhere (for the measure ρ) to their phase-space averages is a consequence of the Birkhoff Ergodic Theorem (see [33], p. 388). Our definition of ergodicity is thus closely related to the classical notion of the equality of phase-space and "time" averages – here time is replaced by the lattice translations. We will show that the ergodic states are the extremal invariant states (and thus that our definition agrees with that of [25], Definition 6.3.1).

It should be noted that to the average physicist or chemist, "macroscopic" might have a different meaning from that used here. Ordinarily one thinks of a "microscopic" size range where quantum effects and thermal fluctuations are important, and a "macroscopic" range which would include the objects of our everyday experience; the exact dividing line is not clearly defined, but atoms are clearly in one range and icebergs in the other. On the other hand, in our mathematical formulation, "macroscopic" refers only to a limit of sizes tending to infinity. In this sense icebergs are still not "macroscopic," and we can easily construct ergodic states with fluctuations on such a scale. This may be worth mentioning, because when we prove that ergodicity is "generic" among invariant states (Lemma IV.3.2) such states will be involved.

After identifying the ergodic states as extremal invariant states, the problem arises of representing an invariant state as a "mixture" of such states, i.e., as the barycenter of a probability measure concentrated on the ergodic states. It will turn out that such a representation exists and is unique. We will also consider a further decomposition into states with short range correlations; this will represent a DLR or KMS state uniquely as a mixture of extremal DLR or KMS states respectively. Although greater generality is possible, we will usually work with the quasi-local algebra of the quantum lattice system. The classical case will always be similar (or easier), and will frequently provide motivation.

Much of this chapter is adapted from [25], Chapter 6, and [29].

IV.2. *Non-commutative ergodic theory*

The classical lattice system can be considered from the point of view of ergodic theory: we have the set Ω with a probability measure ρ and measure-preserving transformations (translations in the lattice). We should try to generalize this theory in dealing with the quantum system. Thus in the classical theory it is useful to consider the Hilbert space $L^2(\rho)$ with the actions of $C(\Omega)$ by multiplication and Z^ν by translation. We obtain an analogous Hilbert space for the quantum system by the *Gel'fand-Naimark-Segal (GNS) construction* as in the following lemma:

LEMMA IV.2.1. *Let* ρ *be any state on* $\mathfrak{A}$.

(i) *There is a* **-representation* π *of* $\mathfrak{A}$ *on a Hilbert space* $\mathcal{H}$ *with a vector* $w \in \mathcal{H}$ *such that* $\pi(\mathfrak{A})w$ *is dense in* $\mathcal{H}$ *and*

$$\rho(A) = \langle w, \pi(A) w \rangle \quad \text{for all } A \in \mathfrak{A}. \tag{3}$$

(ii) *If* ρ *is invariant,* Z^ν *acts on* $\mathcal{H}$ *by unitary operators* $U(i), i \in Z^\nu$, *such that* $U(i)w = w$ *and*

$$\pi(\tau_i A) = U(i)\pi(A)U(i)^{-1} \quad \text{for } A \in \mathfrak{A},\ i \in Z^\nu. \tag{4}$$

(iii) *If* π', $\mathcal{H}'$ *and* w' *also satisfy* (i), *there is a unitary operator* V *from* $\mathcal{H}$ *onto* $\mathcal{H}'$ *such that* $w' = Vw$ *and* $\pi'(A) = V\pi(A)V^{-1}$ *for all* $A \in \mathfrak{A}$. *If* ρ *is invariant and* Z^ν *acts on* $\mathcal{H}'$ *by* $U'(i)$ *as in* (ii), *then* $U'(i) = VU(i)V^{-1}$ *for all* $i \in Z^\nu$.

Proof. Let $\mathfrak{N} = \{A \in \mathfrak{A} : \rho(A^*A) = 0\}$. It is easily seen that $\mathfrak{N}$ is a left ideal of $\mathfrak{A}$, in fact $\mathfrak{N} = \{A : \rho(BA) = 0$ for all $B \in \mathfrak{A}\}$. The quotient algebra $\mathfrak{A}/\mathfrak{N}$ forms a pre-Hilbert space under the inner product

$$<B+\mathfrak{N}, A+\mathfrak{N}> = \rho(B^*A) .$$

Let $\mathcal{H}$ be its completion, and take $w = 1 + \mathfrak{N} \in \mathcal{H}$. The representation π is defined by

$$\pi(A)(B+\mathfrak{N}) = AB + \mathfrak{N} \text{ for } A, B \in \mathfrak{A}$$

since the map $(B+\mathfrak{N}) \mapsto AB + \mathfrak{N}$ on the dense subspace $\mathfrak{A}/\mathfrak{N}$ extends uniquely to a continuous operator on $\mathcal{H}$. Thus $\pi(A)w = A + \mathfrak{N}$. Clearly the requirements of (i) are satisfied.

If ρ is invariant, so is $\mathfrak{N}$. We define the operators $U(i)$ by $U(i)(A+\mathfrak{N}) = \tau_i A + \mathfrak{N}$ for $A \in \mathfrak{A}$, $i \in Z^\nu$. It is easy to verify that the requirements of (ii) are satisfied; for example, (4) is obtained by

$$\begin{aligned}\pi(\tau_i A)(B+\mathfrak{N}) &= (\tau_i A)B + \mathfrak{N} = U(i)(A\tau_{-i}B+\mathfrak{N}) \\ &= U(i)\pi(A)U(i)^{-1}(B+\mathfrak{N}) .\end{aligned}$$

For (iii) we define V by $V\pi(A)w = \pi'(A)w'$ for $A \in \mathfrak{A}$. The requirements of (iii) are easily verified: for example

$$\begin{aligned}VU(i)V^{-1}(\pi'(B)w') &= VU(i)\pi(B)w = V\pi(\tau_i B)w = \pi'(\tau_i B)w' \\ &= U'(i)\pi'(B)w'\end{aligned}$$

for $B \in \mathfrak{A}$, $i \in Z^\nu$. ∎

The classical form of the *mean ergodic theorem* (only slightly modified to replace Z by Z^ν) states that for any $f \in L^2(\rho)$ the translation averages $n^{-\nu} \sum_{i \in C_n} \tau_i f$ tend in L^2 to some translation-invariant function f^*. It is then easily seen that $P\colon f \mapsto f^*$ is the orthogonal projection of $L^2(\rho)$ onto its subspace of translation-invariant functions. This is really just a result about groups of unitary operators on Hilbert spaces:

THEOREM IV.2.2. *Let* $i \mapsto U(i)$ *be a representation of* Z^ν *(as a group under addition) by unitary operators on a Hilbert space* $\mathcal{H}$. *Then for any cubes* C_n *of side* $n \to \infty$, *the averages* $n^{-\nu} \sum_{i \in C_n} U(i)$ *tend in the strong operator topology to the orthogonal projection* P *onto the* U-*invariant vectors of* $\mathcal{H}$.

Proof. It will suffice to prove the case $\nu = 1$, since $n^{-\nu} \sum_{i \in C_n} U(i)$ is the product of 1-dimensional averages of the form $n^{-1} \sum_{i=k+1}^{k+n} U(ie_j)$ where $e_1, \cdots, e_\nu$ are the usual basis of Z^ν, and the projection P is the product of the projections P_j onto the $U(e_j)$-invariant vectors (using the fact that the $U(i)$ commute). Now $U(1)$ has the spectral representation $U(1) = \int_T t\, dP(t)$, where $dP(t)$ is a spectral measure on the unit torus T. We thus obtain

$$n^{-1} \sum_{i=k+1}^{k+n} U(i) = \int_T f_n(t)\, dP(t)$$

where

$$f_n(t) = n^{-1} \sum_{i=k+1}^{k+n} t^i = \begin{cases} 1 & \text{for } t = 1 \\ \dfrac{t^{k+n+1} - t^{k+1}}{n(t-1)} & \text{for } t \neq 1 . \end{cases}$$

Note that $|f_n(t)| \leq 1$ for all $t \in T$, and f_n tends pointwise on T to the function which is 1 at $t=1$ and 0 elsewhere. Therefore $\int_T f_n(t)\, dP(t)$ tends strongly to $P\{1\}$, which is the projection P on the U-invariant vectors of $\mathcal{H}$. ∎

In the classical case $\mathcal{H}$ is $L^2(\rho)$ for an invariant state ρ and $U(i)$ is translation by $i \in Z^\nu$. In the quantum case we use the GNS construction, obtaining the following results:

COROLLARY IV.2.3. *Let* ρ *be any invariant state of* $\mathfrak{A}$. *Then for any* $A \in \mathfrak{A}$,

$$\lim_{n\to\infty} \rho(c_n(A)^* c_n(A)) = \|P\pi(A)w\|^2 \geq |\rho(A)|^2 . \tag{5}$$

Thus ρ *is ergodic if and only if* $P\mathcal{H}$ *consists of scalar multiples of* w. *If* $\rho = \frac{1}{2}\rho_1 + \frac{1}{2}\rho_2$ *for invariant states* ρ_1, ρ_2 *which differ on* $A \in \mathfrak{A}$, *then* $\lim_{n\to\infty} \rho(c_n(A)^* c_n(A)) > |\rho(A)|^2$. *Thus ergodic states are extreme points of* E^I.

Proof. Formula (5) follows directly from the theorem, using

$$\rho(c_n(A)^* c_n(A)) = \|\pi(c_n(A))w\|^2 = \left\| n^{-\nu} \sum_{i \in C_n} U(i)\pi(A)w \right\|^2$$

and

$$\|P\pi(A)w\| \geq |\langle w, P\pi(A)w\rangle| = |\rho(A)| \quad (\text{note } \|w\| = 1) .$$

Equality holds in (5) if and only if $P\pi(A)w$ is a scalar multiple of w; since $\pi(\mathfrak{A})w$ is dense in $\mathcal{H}$, $P\mathcal{H}$ contains only these vectors if equality holds for all A. In defining ergodicity by equation (2) we considered only self-adjoint A, so in general we must take real and imaginary parts.

If ρ_1 and ρ_2 are as above, we have

$$\begin{aligned} |\rho(A)|^2 &= |\tfrac{1}{2}\rho_1(A) + \tfrac{1}{2}\rho_2(A)|^2 \\ &< \tfrac{1}{2}|\rho_1(A)|^2 + \tfrac{1}{2}|\rho_2(A)|^2 \leq \lim_{n\to\infty} \rho(c_n(A)^* c_n(A)) \end{aligned}$$

so that ρ is not ergodic. ∎

In the classical case this corollary states that ρ is ergodic if and only if all invariant functions in $L^2(\rho)$ are constant almost everywhere. From this, we can show that extreme points of E^I are ergodic. For if the invariant state ρ is not ergodic, there are functions in $L^2(\rho)$ which are not almost everywhere constant. The inverse image of some open set in $\mathbb{C}$ under such a function is then a translation-invariant measurable subset S of Ω with $0 < \rho(S) < 1$. We can write $\rho = \rho(S)\rho_1 + (1-\rho(S))\rho_2$ where $\rho_1 = \rho(S)^{-1}\chi_S\rho$ and $\rho_2 = (1-\rho(S))^{-1}(1-\chi_S)\rho$ are distinct invariant states (χ_S being the characteristic function of S). Thus ρ is not extremal.

In the quantum system, the analogue of S (or more precisely, of the operator of multiplication by χ_S on $L^2(\rho)$) would be an orthogonal projection E in $\mathcal{H}$ which commutes with both $\pi(\mathfrak{A})$ and all the $U(i)$, but is not 1 or 0. Given such a projection, we can show that ρ is not extremal. It will be useful to state this in the language of von Neumann algebras. We define

$$\mathfrak{R} = (\pi(\mathfrak{A}) \cup U(Z^\nu))'' \tag{6}$$

the smallest von Neumann algebra on $\mathcal{H}$ containing all the operators $\pi(A)$ and $U(i)$. A self-adjoint set F of operators is *irreducible* if its commutant F', the von Neumann algebra of operators commuting with all elements of F, contains only multiples of the identity. This is equivalent to the condition that F' contains no nontrivial projection, i.e., $\mathcal{H}$ has no nontrivial F-invariant closed subspaces.

LEMMA IV.2.4. *The invariant state* ρ *is extremal in* E^I *if and only if* $\mathfrak{R}$ *is irreducible.*

Proof. If $\mathfrak{R}$ is not irreducible, there is a nontrivial projection E in $\mathfrak{R}'$ (when we say "projection" in the context of von Neumann algebras, we will always mean orthogonal projection). Thus E commutes with all $\pi(A)$ and $U(i)$, so that invariant states ρ_1, ρ_2 are defined by

$$\rho_1(A) = \frac{\langle w, E\pi(A)w\rangle}{\langle w,Ew\rangle} \qquad \rho_2(A) = \frac{\langle w,(1-E)\pi(A)w\rangle}{\langle w,(1-E)w\rangle} .$$

Since $\rho = \langle w,Ew\rangle \rho_1 + (1-\langle w,Ew\rangle)\rho_2$, it is not extremal.

Conversely, suppose $\rho = \frac{1}{2}\rho_1 + \frac{1}{2}\rho_2$ *for distinct invariant states* ρ_1, ρ_2. Then we can define a sesquilinear form on $\pi(\mathfrak{A})w$ by $Q(\pi(A)w, \pi(B)w) = \rho_1(A^*B)$. Since $\rho_1(A^*A) \leq 2\rho(A^*A) = 2\|\pi(A)w\|^2$ this extends uniquely to a continuous form on $\mathcal{H}$. Therefore $Q(u,v) = \langle u,Tv\rangle$ for some bounded operator T on $\mathcal{H}$. Now T commutes with each $\pi(A)$ since for any $A, B, C \in \mathfrak{A}$

$$\langle \pi(B)w, T\pi(A)\pi(C)w\rangle = \rho_1(B^*AC) = \langle \pi(A)^*\pi(B)w, T\pi(C)w\rangle$$
$$= \langle \pi(B)w, \pi(A)T\pi(C)w\rangle .$$

Moreover, T commutes with each $U(i)$ since

$$\langle \pi(A)w, TU(i)\pi(B)w\rangle = \rho_1(A^*\tau_i B) = \rho_1(\tau_{-i}(A)^*B)$$
$$= \langle U(-i)\pi(A)w, T\pi(B)w\rangle = \langle \pi(A)w, U(i)T\pi(B)w\rangle .$$

Thus $T \in \pi(\mathfrak{A})' \cap U(Z^\nu)' = \mathfrak{R}'$. Since T is not a multiple of 1, $\mathfrak{R}$ is not irreducible. ∎

So far, everything we have done is applicable in an extremely general setting: a C^*-algebra with identity, an action of Z^ν by automorphisms of the algebra, and an invariant state. It is even possible to replace Z^ν by any topological group G with an invariant mean, the action on the C^*-algebra being strongly continuous (see [25], Chapter 6). If we want extreme points of E^I to be ergodic, some further properties of our system must be used. Otherwise one could take the C^*-algebra of complex $n \times n$ matrices, with the group acting trivially $(U(i) = 1)$. Then all vector states will be extremal, but none will be ergodic. The precise condition we need is that $\mathfrak{A}$ is G-*abelian* (see [25], Definition 6.2.6): roughly speaking, when two elements of $\mathfrak{A}$ are translated far enough apart, they "almost" commute.

For any space $\mathfrak{F}$ of operators on $\mathcal{H}$, $P\mathfrak{F}P$ will denote the space of operators $\{PT\restriction_{P\mathcal{H}}: T \in \mathfrak{F}\}$ on the Hilbert space $P\mathcal{H}$.

LEMMA IV.2.5. *Let* $\mathfrak{M}$ *be any von Neumann algebra on* $\mathcal{H}$ *containing the projection* P. *Then* $P\mathfrak{M}P$ *is a von Neumann algebra on* $P\mathcal{H}$ *and* $(P\mathfrak{M}P)' = P\mathfrak{M}'P$.

Proof. First we show $P\mathfrak{M}P$ is strongly closed. If an operator B on $P\mathcal{H}$ is the strong limit of a net $PA_a\restriction_{P\mathcal{H}}$ with $A_a \in \mathfrak{M}$, then the operators PA_aP on $\mathcal{H}$ converge strongly to BP, so $BP \in \mathfrak{M}$ and $B = PBP\restriction_{P\mathcal{H}} \in P\mathfrak{M}P$. The other requirements for a von Neumann algebra are clearly satisfied.

For any $A \in \mathfrak{M}$ and $B \in \mathfrak{M}'$ we have $PAPBP = PABP = PBAP = PBPAP$ so $PA\restriction_{P\mathcal{H}}$ and $PB\restriction_{P\mathcal{H}}$ commute. Thus $P\mathfrak{M}'P \subseteq (P\mathfrak{M}P)'$. Conversely, suppose $T \in (P\mathfrak{M}P)'$. Since every member of a von Neumann algebra is a linear combination of four unitary elements of the von Neumann algebra, we can assume T is unitary. For any $u_1, \cdots, u_n \in P\mathcal{H}$ and $A_1, \cdots, A_n \in \mathfrak{M}$ we have

$$\Big\|\sum_i A_i T u_i\Big\|^2 = \sum_{i,j} \langle A_i T U_i, A_j T U_j \rangle = \sum_{i,j} \langle P A_j^* A_i T U_i, T u_j \rangle$$
$$= \sum_{i,j} \langle T P A_j^* A_i u_i, T u_j \rangle = \Big\|\sum_i A_i u_i\Big\|^2 .$$

Let $\mathcal{K}$ be the closed linear span of $\mathfrak{M}P\mathcal{H}$ in $\mathcal{H}$. The isometry $\sum_i A_i u_i \mapsto \sum_i A_i T u_i$ extends uniquely to a bounded linear operator on $\mathcal{K}$, and then to an operator S on $\mathcal{H}$ by taking S to be zero on $\mathcal{K}^\perp$. Then $S \in \mathfrak{M}'$, for if $A \in \mathfrak{M}$,

$$SA\Big(\sum_i A_i u_i\Big) = \sum_i AA_i T u_i = AS\Big(\sum_i A_i u_i\Big) \quad \text{for } A_i \in \mathfrak{M},\ u_i \in P\mathcal{H}$$

while $SAv = 0 = ASv$ for $v \in \mathcal{K}^\perp$. Moreover for $u \in P\mathcal{H}$

$$PSu = PTu = Tu$$

so $T = PS\restriction_{P\mathcal{H}} \in P\mathfrak{M}'P$. ■

LEMMA IV.2.6. $P\mathfrak{R}P = (P\pi(\mathfrak{A})P)''$ *and is abelian.*

Proof. Consider the subspace of $\mathfrak{R}$ consisting of finite sums $\sum_{i\in\Lambda} U(i)\pi(A_i)$ with $A_i \in \mathfrak{A}$. Since $U(i)\pi(A) = \pi(\tau_i A)U(-i)$, this subspace forms a *-subalgebra containing all the $\pi(A)$ and $U(i)$. Therefore its strong closure coincides with $\mathfrak{R}$. So for any $T \in \mathfrak{R}$, there is a net of operators T_a of the form $\sum_{i\in\Lambda} U(i)\pi(A_i)$ converging strongly to T. Then $PT\restriction_{P\mathcal{H}}$ is the strong limit of the operators $PT_a\restriction_{P\mathcal{H}}$ of the form $\sum_{i\in\Lambda} PU(i)\pi(A_i)\restriction_{P\mathcal{H}}$ $= P\pi(\sum_{i\in\Lambda} A_i)\restriction_{P\mathcal{H}}$ in $P\pi(\mathfrak{A})P$, so that $P\mathfrak{R}P \subseteq (P\pi(\mathfrak{A})P)''$. On the other hand, $P\pi(\mathfrak{A})P \subseteq P\mathfrak{R}P$, which is a von Neumann algebra since $P \in \mathfrak{R}$ by the mean ergodic theorem. Thus $P\mathfrak{R}P = (P\pi(\mathfrak{A})P)''$.

Now we show $(P\pi(\mathfrak{A})P)''$ is abelian. Suppose A and B are in some $\mathfrak{A}_\Lambda$ with Λ finite. Then there are at most $|\Lambda|^2$ lattice points i for which A and $\tau_i B$ do not commute, since this requires $\Lambda \cap (i+\Lambda) \neq \emptyset$. Thus the commutator $[A, c_n(B)]$ has norm at most $2n^{-1}|\Lambda|^2\|A\|\,\|B\|$ and tends to 0 as $n\to\infty$. Thus

$$\begin{aligned} P\pi(A)P\pi(B)\restriction_{P\mathcal{H}} &= \text{s-lim}\, P\pi(A)\pi(c_n(B))\restriction_{P\mathcal{H}} \\ &= \text{s-lim}\, P\pi(c_n(B))\pi(A)\restriction_{P\mathcal{H}} = P\pi(B)P\pi(A)\restriction_{P\mathcal{H}}. \end{aligned}$$

By continuity this equality extends to all $A, B \in \mathfrak{A}$. Thus $P\pi(\mathfrak{A})P$ is abelian, and hence $(P\pi(\mathfrak{A})P)''$ is also abelian. ■

THEOREM IV.2.7. *Extreme points of* E^I *are ergodic.*

Proof. If ρ is extreme, $\mathfrak{R}$ is irreducible by Lemma IV.2.4. Then $P\mathfrak{R}P$ is irreducible by Lemma IV.2.5. By Lemma IV.2.6, $P\mathfrak{R}P$ is abelian, so $P\mathfrak{R}P \subseteq (P\mathfrak{R}P)'$ contains only multiples of 1. But then $P\mathcal{H} = \mathbf{C}w$, so ρ is ergodic. ■

IV.3. *Integral representations*

We now want to show that any invariant state can be uniquely represented by a probability measure on the ergodic states. The existence of this measure follows from *Choquet's Theorem* ([34], p. 19): *If* K *is a metrizable compact convex set and* $x \in K$, *there is a probability measure* μ *with barycenter* x (i.e., $\mu(f) = f(x)$ for all continuous affine functions f on K) *concentrated on the set* $\mathcal{E}(K)$ *of extreme points of* K (i.e., $\mu(K \setminus \mathcal{E}(K)) = 0$). The uniqueness still remains to be shown; one method will be to show E^I is a *Choquet simplex*.

The importance of the metrizability of K in the above theorem is that this implies that the extreme points of K form a Baire set, in fact a G_δ. For non-metrizable K the extreme points need not even form a Borel set, and various complications are encountered: see [34], Section 4. Fortunately, our algebras $\mathfrak{A}$ and $C(\Omega)$ are separable, so their spaces of states are metrizable (in the weak-* topology, as usual).

LEMMA IV.3.1. *If* K *is a metrizable compact convex set, then* $\mathcal{E}(K)$ *is a* G_δ.

Proof. Let d be a metric on K. For each n let F_n be the set of points of K which can be written as $\frac{x}{2} + \frac{y}{2}$ with $x,y \in K$ and $d(x,y) \geq \frac{1}{n}$. Then F_n is a continuous image of the compact set of pairs (x,y) in $K \times K$ with $d(x,y) \geq \frac{1}{n}$, so F_n is compact. Thus $\mathcal{E}(K)$, which is the intersection of the complements of the F_n, is a G_δ. ∎

We make a slight detour at this point, to show that for our lattice systems, the ergodic states are actually a *dense* G_δ. Thus we have the rather counter-intuitive situation of a convex set in which "generic" points are extremal. Of course, this is for the weak-* topology; if we took instead the norm topology on the invariant states, we would find the ergodic states forming a closed nowhere-dense set (in fact the distance between any two ergodic states is 2: see Corollary IV.4.2). For eventual use in Theorem V.2.2, we refine our result further by considering the mean entropy.

LEMMA IV.3.2. *Ergodic states are dense in* E^I. *Moreover, for any* $r \in \mathbf{R}$, *ergodic states are dense in the invariant states with mean entropy* $s(\rho) > r$.

Proof. Given an invariant state ρ_0 and $A_1, \cdots, A_n \in \mathfrak{A}$, we want to find an ergodic state ρ with $|\rho(A_i) - \rho_0(A_i)| < 1$ and (if $s(\rho_0) > r$) $s(\rho) > r$. We can assume without loss of generality that all the A_i are in some $\mathfrak{A}_\Lambda$ with Λ finite. Let C be a large cube containing Λ. We cover $\mathbf{Z}^\nu$ by a regular lattice of disjoint translates of C, and take the product state ρ_1 of the restrictions of ρ to each of these cubes. Thus the state is unchanged inside each cube, but the cubes have been made independent. Our state ρ_1 is periodic, but not invariant. Therefore we let $\rho = |C|^{-1} \sum_{j \in C} \rho_1 \circ \tau_j$, which is invariant. The difference $\rho(A_i) - \rho_0(A_i)$ is due to those $j \in C$ such that $j + \Lambda$ intersects more than one of our cubes. For C sufficiently large, these j will constitute such a small fraction of C that we will have $|\rho(A_i) - \rho_0(A_i)| < 1$. The periodic state ρ_1 is easily seen to have mean entropy $s(\rho_1) = |C|^{-1} S_C(\rho_0)$. Since s is affine (on periodic states as well as invariant ones), $s(\rho) = |C|^{-1} S_C(\rho_0)$ as well. If $s(\rho_0) > r$, by taking C sufficiently large we can make $s(\rho) > r$. To see that ρ is ergodic, note that if $A \in \mathfrak{A}_{\Lambda'}$ with Λ' finite, there is some N such that $\rho_1(\tau_i(A)^* \tau_j(A)) = \rho_1(\tau_i A^*) \rho_1(\tau_j A)$ for $|i-j| > N$. Thus it is easily seen that

$$\lim_{n\to\infty} \rho_1(c_n(A)^* c_n(A)) = |C|^{-2} \left| \sum_{i \in C} \rho_1(\tau_i A) \right|^2 = |\rho(A)|^2 .$$

Also, $\lim_{n\to\infty} \rho(c_n(A)^* c_n(A)) = |\rho(A)|^2$, and this extends by continuity to any $A \in \mathfrak{A}$. ∎

This lemma means that a finite number of measurements of quasi-local observables, and even the mean entropy, to within a finite range of accuracy can never prove that a state is nonergodic. Of course, this should not discourage experimentalists!

THEOREM IV.3.3. *For any* $\rho \in E^I$, *there is a unique probability measure* μ *concentrated on* $\mathcal{E}(E^I)$ *with barycenter* ρ. *If* ρ *is an invariant equilibrium state for* $\Phi \in \mathfrak{B}$, *then* μ *is concentrated on extremal invariant equilibrium states for* Φ.

Proof. Since Choquet's Theorem provides us with existence of μ (and if ρ is an invariant equilibrium state we can apply Choquet's Theorem to the convex compact set of invariant equilibrium states for Φ), we need only prove uniqueness. For $A \in \mathfrak{A}$ let $\hat{A}$ be the continuous function $\rho' \mapsto \rho'(A)$ on E^I. By the Stone-Weierstrass Theorem the algebra generated by the $\hat{A}$ is dense in $C(E^I)$, so to determine a measure uniquely we need only give its values on products $\hat{A}_1\hat{A}_2 \cdots \hat{A}_k$ of these functions. For $\rho' \in E^I$ (and using the subscript ρ' to distinguish the GNS construction and related objects for ρ' from those for other invariant states) we have

$$\pi_{\rho'}(c_n(A))|_{P_{\rho'}\mathcal{H}_{\rho'}} = n^{-\nu} \sum_{i \in C_n} U_{\rho'}(i)\pi_{\rho'}(A)|_{P_{\rho'}\mathcal{H}_{\rho'}} \xrightarrow{s} P_{\rho'}\pi_{\rho'}(A)|_{P_{\rho'}\mathcal{H}_{\rho'}}.$$

Since $\|c_n(A)\| \leq \|A\|$ this implies

$$\pi_{\rho'}(c_n(A_1)\cdots c_n(A_k))|_{P_{\rho'}\mathcal{H}_{\rho'}} \xrightarrow{s} P_{\rho'}\pi_{\rho'}(A_1)P_{\rho'}\cdots P_{\rho'}\pi_{\rho'}(A_k)|_{P_{\rho'}\mathcal{H}_{\rho'}}$$

$$(7) \qquad \rho'(c_n(A_1)\cdots c_n(A_k)) \to \langle w_{\rho'}, \pi_{\rho'}(A_1)P_{\rho'}\pi_{\rho'}(A_2)P_{\rho'}\cdots P_{\rho'}\pi_{\rho'}(A_k)w_{\rho'}\rangle .$$

If ρ' is ergodic, $P_{\rho'}$ is just the one-dimensional projection on $w_{\rho'}$, so the right-hand side of (7) is $\rho'(A_1)\cdots\rho'(A_k) = (\hat{A}_1\hat{A}_2\cdots\hat{A}_k)(\rho')$. Thus if μ is a probability measure concentrated on $\mathcal{E}(E^I)$ with barycenter ρ, we have

$$(8) \qquad \begin{aligned} \mu(\hat{A}_1\hat{A}_2\cdots\hat{A}_k) &= \int_{\mathcal{E}(E^I)} \lim_{n\to\infty} \rho'(c_n(A_1)\cdots c_n(A_k))\mu(d\rho') \\ &= \lim_{n\to\infty} \rho(c_n(A_1)\cdots c_n(A_k)) = \langle w_\rho, \pi_\rho(A_1)P_\rho\pi_\rho(A_2)\cdots P_\rho\pi_\rho(A_k)w_\rho\rangle . \end{aligned}$$ ∎

Let K be a compact convex set in a locally convex topological vector space X. We can assume (if necessary adding a dimension to X) that K is contained in a hyperplane not passing through 0. The *cone through* K is the set $C = \{tx: 0 \leq t < \infty, x \in K\}$. This defines a partial order on X by $x \geq y$ if $x-y \in C$. We call K a *Choquet simplex* if C is a *lattice* in this ordering, i.e., if any $x,y \in C$ have a least upper bound $x \vee y$ and a greatest lower bound $x \wedge y$. For finite-dimensional sets, this definition turns out to be equivalent to the usual definition of a simplex as the convex hull of a finite number of affinely independent points. We then have the theorem of Choquet and Meyer (see [34], p. 66):

A compact convex metrizable set K *is a Choquet simplex if and only if each point of* K *is the barycenter of a unique probability measure concentrated on* $\mathfrak{E}(K)$.

This theorem, together with Theorem IV.3.3, shows that the set E^I of invariant states forms a Choquet simplex. However, it is more instructive to prove this directly. As usual, the classical case is easiest. It clearly suffices to prove that for any ρ in the cone C (of invariant positive finite measures on Ω), the subset $C_\rho = \{\rho' \in C: \rho' \leq \rho\}$ is a lattice. Now the Radon-Nikodym Theorem provides an order-preserving 1-1 map of C_ρ into $\{f \in L^\infty(\rho): 0 \leq f \leq 1\}$. If f is translation-invariant, so is the measure $f\rho$; conversely, if $f\rho$ is invariant, the uniqueness of the Radon-Nikodym derivative shows that f is invariant (up to sets of ρ-measure zero, of course). Thus the image of C_ρ under this map is the lattice $\{f \in L^\infty(\rho)$ translation-invariant: $0 \leq f \leq 1\}$. Therefore C_ρ is also a lattice, as required.

Consider now $C_\rho = \{\rho' \in C: \rho' \leq \rho\}$ in the quantum case (here C is the cone of invariant positive linear functionals on $\mathfrak{A}$). We can assume for simplicity $\rho \in E^I$, and as in Section IV.2 take the GNS construction, the algebra $\mathfrak{R}$, etc., for ρ. Now as in the proof of Lemma IV.2.4, we find that for any $\rho' \in C_\rho$ there is an operator $T \in \mathfrak{R}'$ such that

$$\rho'(A) = \langle w, T\pi(A)w \rangle \quad \text{for } A \in \mathfrak{A}. \tag{9}$$

Moreover, $0 \le T \le 1$ since $0 \le \rho'(A^*A) = <\pi(A)w, T\pi(A)w> \le \rho(A^*A)$ for all $A \in \mathfrak{A}$. Conversely, if $T \in \mathfrak{R}'$ with $0 \le T \le 1$ it is easily seen that (9) defines an invariant positive linear functional ρ' on $\mathfrak{A}$ with $\rho' \le \rho$. Thus we have a 1-1 order-preserving map of C_ρ onto $\{T \in \mathfrak{R}': 0 \le T \le 1\}$, the analogue of the map given by the Radon-Nikodym derivative in the classical case. To complete the proof, we must show that the self-adjoint elements of $\mathfrak{R}'$ form a lattice. This will be true if $\mathfrak{R}'$ is abelian: any abelian C^*-algebra is *-isomorphic (via the Gel'fand transform) to $C(M)$, where M is its maximal ideal space; and real-valued functions in $C(M)$ form a lattice, taking inf and sup pointwise.

LEMMA IV.3.4. *The algebra* $\mathfrak{R}'$ *is abelian, and* $r: C \mapsto C\restriction_{P\mathcal{H}}$ *is a* **-isomorphism of* $\mathfrak{R}'$ *onto* $P\mathfrak{R}P$.

Proof. Clearly r is a *-homomorphism onto $P\mathfrak{R}'P$. It is 1-1 since if $Cw = 0$ with $C \in \mathfrak{R}'$ we have $C\pi(A)w = \pi(A)Cw = 0$ for all $A \in \mathfrak{A}$. Now by Lemma IV.2.5, $P\mathfrak{R}'P = (P\mathfrak{R}P)'$. The von Neumann algebra $P\mathfrak{R}P$ is abelian by Lemma IV.2.6, and has the cyclic vector $w \in P\mathcal{H}$. The next lemma will show that this implies $P\mathfrak{R}P = (P\mathfrak{R}P)' = r(\mathfrak{R}')$; thus $\mathfrak{R}'$, being isomorphic to an abelian algebra, is also abelian.

LEMMA IV.3.5. *An abelian von Neumann algebra* $\mathfrak{M}$ *with a cyclic vector* w *is maximal abelian, i.e.,* $\mathfrak{M}' = \mathfrak{M}$.

Proof. It suffices to show that $\mathfrak{M}$ contains every self-adjoint element B of $\mathfrak{M}'$. Let $A_n \in \mathfrak{M}$ with $A_n w \to Bw$. Since $\mathfrak{M}$ is abelian and B is self-adjoint, $A_n - B$ is normal (commutes with $A_n^* - B$), and so $\|A_n^* w - Bw\| = \|A_n w - Bw\| \to 0$. Thus for any $C \in \mathfrak{M}'$ and $A, A' \in \mathfrak{M}$

$$<Aw, BCA'w> = <ABw, CA'w> = \lim_{n\to\infty} <AA_n^* w, CA'w>$$

$$= \lim_{n\to\infty} <Aw, CA'A_n w> = <Aw, CA'Bw> = <Aw, CBA'w>$$

so $B \in \mathfrak{M}'' = \mathfrak{M}$. ∎

The ergodic decomposition of Theorem IV.3.3 is a special case of a general theory of decompositions of states (see [27]). Given a state ρ of a C*-algebra $\mathfrak{A}$, one considers instead of $\mathfrak{R}'$ an arbitrary abelian von Neumann algebra $\mathfrak{B} \subset \pi(\mathfrak{A})'$, with the projection $P_{\mathfrak{B}}$ on the closure of $\mathfrak{B}w$ in $\mathcal{H}$. One then obtains a probability measure μ on the states of $\mathfrak{A}$ with barycenter ρ by

$$\mu(\hat{A}_1 \cdots \hat{A}_k) = \langle w, \pi(A_1) P_{\mathfrak{B}} \pi(A_2) \cdots P_{\mathfrak{B}} \pi(A_k) w \rangle \quad \text{for } A_1, \cdots, A_k \in \mathfrak{A} .$$

However, we will only be interested in one more case of this theory, where $\mathfrak{B}$ is the *algebra at infinity* (see below), and only when ρ is a KMS or DLR state; the measure μ then turns out to be concentrated on the extremal KMS or DLR states. Instead of obtaining the measure as above, we will show that the KMS or DLR states (for a given interaction) form a Choquet simplex.

For each finite $\Lambda \subset Z^\nu$, $\mathfrak{A}_{\Lambda^c}$ is the closure of $\bigcup_{\substack{\Lambda' \cap \Lambda = \emptyset \\ \Lambda' \text{ finite}}} \mathfrak{A}_{\Lambda'}$ (in the classical case, this is $C(\Omega_{\Lambda^c})$). Given a state ρ, the *algebra at infinity* for ρ is the von Neumann algebra $\mathfrak{B}_\rho = \bigcap_{\Lambda \text{ finite}} \pi(\mathfrak{A}_{\Lambda^c})''$ (where π is the representation in the GNS construction for ρ). In the classical case, $\pi(C(\Omega_{\Lambda^c}))''$ consists of all functions in $L^\infty(\Omega_{\Lambda^c}, \rho)$, considered as multiplication operators on $L^2(\Omega, \rho)$. Thus $\mathfrak{B}_\rho$ consists of all functions $f \in L^\infty(\rho)$ such that for any finite Λ, f is equal almost everywhere to some function that does not depend on the configuration in Λ. For example, in the Ising model we could take

$$f(\omega) = \liminf_{n \to \infty} (c_n(\sigma_0))(\omega) \quad \text{(for a given sequence of cubes } C_n) .$$

This function will be in $\mathfrak{B}_\rho$ for every ρ, but for many of them it will be equal to a constant almost everywhere.

LEMMA IV.3.6. *For the quantum lattice system,* $\mathfrak{B}_\rho$ *is the center of* $\pi(\mathfrak{A})''$.

Proof. For any $B \in \mathfrak{B}_\rho$ and any finite Λ, $B \in \pi(\mathfrak{A}_{\Lambda^c})''$ commutes with each element of $\pi(\mathfrak{A}_\Lambda)$ since members of $\mathfrak{A}_{\Lambda^c}$ and $\mathfrak{A}_\Lambda$ commute with each other. By continuity, B commutes with each element of $\pi(\mathfrak{A})$, thus is in the center $\pi(\mathfrak{A})' \cap \pi(\mathfrak{A})''$ of $\pi(\mathfrak{A})''$. Conversely, suppose $\Lambda \subset \mathbf{Z}^\nu$ is finite and $B \in \pi(\mathfrak{A})' \cap \pi(\mathfrak{A})''$. Then B is the strong limit of a net $\pi(B_\alpha)$ with $B_\alpha \in \mathfrak{A}$. Take an orthonormal basis $\{u_1, \cdots, u_n\}$ of $\mathcal{H}_\Lambda$ and corresponding matrix units $E_{ij} \in \mathfrak{A}_\Lambda (E_{ij}u_k = \delta_{jk}u_i)$. Note that $\sum_{i,j} E_{ij}E_{ji} = n \sum_i E_{ii} = n1$, and $\mathrm{tr}_\Lambda C = n^{-1} \sum_{i,j} E_{ij}CE_{ji}$ for any $C \in \mathfrak{A}$. Thus $B = n^{-1} \sum_{i,j} \pi(E_{ij}E_{ji})B = n^{-1} \sum_{i,j} \pi(E_{ij})B\,\pi(E_{ji}) = \text{s-lim}\, \pi(\mathrm{tr}_\Lambda B_\alpha)$. Since $\mathrm{tr}_\Lambda B_\alpha \in \mathfrak{A}_{\Lambda^c}$, we have $B \in \pi(\mathfrak{A}_{\Lambda^c})''$. ∎

A state ρ of $\mathfrak{A}$ will be said to have *short range correlations* if for every $A \in \mathfrak{A}$ there is $\Lambda \subset \mathbf{Z}^\nu$ finite such that

$$(10) \qquad |\rho(AB) - \rho(A)\rho(B)| \leq \|B\| \quad \text{for all } B \in \mathfrak{A}_{\Lambda^c}.$$

By multiplying A by a scalar, we can equivalently replace the right-hand side of (10) by $\varepsilon\|B\|$ for any $\varepsilon > 0$ (of course Λ will depend on ε). For invariant states, this represents a stronger property than ergodicity (implying that $\rho(A\tau_i B)$ converges to $\rho(A)\rho(B)$, not just in the Cesaro sense; in ergodic theory this property is called *mixing*). In fact, for invariant states of classical one-dimensional systems, the short range correlations property implies that the system is a *K-system* (see [18]).

LEMMA IV.3.7. *The state ρ has short range correlations if and only if its algebra at infinity $\mathfrak{B}_\rho = \mathbf{C}1$.*

Proof. Suppose ρ does not have short range correlations. Let Λ_n be a sequence of finite subsets of $\mathbf{Z}^\nu$ increasing to infinity (i.e., eventually containing each site). Then for some $A \in \mathfrak{A}$ there is a sequence $B_n \in \mathfrak{A}_{\Lambda_n^c}$ with $|\rho(AB_n) - \rho(A)\rho(B_n)| > \|B_n\| = 1$. Now the unit ball of a von

Neumann algebra is easily seen to be compact in the weak operator topology, so the sequence $\pi(B_n)$ has a weak limit point B, which must be in each $\pi(\mathfrak{A}_{\Lambda_n^c})''$ and thus in $\mathfrak{B}_\rho$. Moreover

$$|<w, \pi(A)Bw> - <w, \pi(A)w><w,Bw>| \geq \lim_{n \to \infty} \inf |\rho(AB_n) - \rho(A)\rho(B_n)| \geq 1$$

so B is not a scalar multiple of 1.

   Conversely, suppose $B \epsilon \mathfrak{B}_\rho$, $A \epsilon \mathfrak{A}$ and take Λ so that (10) holds. By the Kaplansky Density Theorem there is a net $\pi(A_\alpha)$ converging strongly to B with $A \epsilon \mathfrak{A}_{\Lambda^c}$ and $\|\pi(A_\alpha)\| \leq \|B\|$. The lemma below shows that we can take $\|A_\alpha\| \leq \|B\|$. Thus

$$|<w, \pi(A)Bw> - <w, \pi(A)w><w,Bw>| = \lim_\alpha |\rho(AA_\alpha) - \rho(A)\rho(A_\alpha)| \leq \|B\| .$$

If ρ has short range correlations, by homogeneity in A the left-hand side of the above expression is actually 0. Thus we obtain $Bw = <w,Bw>w$, and (since $B \epsilon \pi(\mathfrak{A})'$) $B = <w,Bw>1$.   ∎

LEMMA IV.3.8. *Let* $\pi: \mathfrak{A} \to \mathfrak{B}$ *be a* *-*homomorphism of* C^*-*algebras. Then for any* $B \epsilon \pi(\mathfrak{A})$ *there is* $A \epsilon \mathfrak{A}$ *with* $\pi(A) = B$ *and* $\|A\| \leq \|B\|$.

Proof. Let f be the continuous function $f(t) = \inf\{1, \|B\| t^{-\frac{1}{2}}\}$ on $[0, \infty)$. If $B = \pi(A_0)$ let $A = A_0 f(A_0^* A_0) \epsilon \mathfrak{A}$ (if $\mathfrak{A}$ does not have an identity, we mean $A_0 - A_0(1-f)(A_0^* A_0)$). Then $\pi(f(A_0^* A_0)) = f(B^* B) = 1$ so $\pi(A) = B$. Moreover, since $tf(t)^2 \leq \|B\|^2$,

$$\|A\|^2 = \|f(A_0^* A_0)A_0^* A_0 f(A_0^* A_0)\| \leq \|B\|^2 .   ∎$$

   Consider now the classical lattice system, with a fixed interaction $\Phi \epsilon \tilde{\mathfrak{B}}$. Let Γ^Φ be the set of DLR states for Φ. Since these can be characterized by the equations (6) of Section III.2, Γ^Φ is convex and compact (in the weak-* topology on states).

LEMMA IV.3.9. *A state* $\rho \in \Gamma^{\Phi}$ *is extremal in* Γ^{Φ} *if and only if it has short range correlations.*

Proof. If ρ does not have short range correlations, there is $h \in \mathfrak{B}_{\rho}$ non-constant with $0 \leq h \leq 1$. The state $\rho' = \rho(h)^{-1}h\rho$ will then satisfy the DLR equations: for any finite Λ, there is $\tilde{h} \in L^{\infty}(\Omega_{\Lambda^c}, \rho)$ with $h = \tilde{h}$ ρ-almost everywhere. Taking $\rho'(s, d\tau) = \tilde{h}(\tau)\rho(s, d\tau)$, it is clear that ρ' satisfies the DLR equations (formula (3) of Section III.1). Then ρ is a convex combination of the DLR states ρ' and $\rho(1-h)^{-1}(1-h)\rho$. Conversely, if ρ is a nontrivial convex combination of distinct DLR states ρ', ρ'', then ρ' is absolutely continuous with respect to ρ, so $\rho' = h\rho$ with h non-constant and bounded (since ρ' is less than some multiple of ρ). Now for the discrete-spin case it is clear that $\rho'(s, d\tau) = h(s \times \tau)\rho(s, d\tau)$ for $s \in \Omega_{\Lambda}$, $\tau \in \Omega_{\Lambda^c}$, but in the continuous case we must work a little. For $f \in C(\Omega_{\Lambda})$ and $g \in C(\Omega_{\Lambda^c})$ (Λ fixed and finite) we have

$$\int_{\Omega_{\Lambda}} f(s)\rho'(s,g)\mu_0(ds) = \rho'(f \otimes g) = \int_{\Omega_{\Lambda}} \int_{\Omega_{\Lambda^c}} h(s \times \tau) f(s) g(\tau) \rho(s, d\tau) \mu_0(ds)$$

which implies $\rho'(s,g) = \int_{\Omega_{\Lambda^c}} h(s \times \tau) g(\tau) \rho(s, d\tau)$ for μ_0-almost every $s \in \Omega_{\Lambda}$. Now both sides of this expression are continuous in g; since $C(\Omega_{\Lambda})$ is separable, there is a set of μ_0-measure one in Ω_{Λ} within which the equality holds for every g. Choose a point s_0 in this set, and let $\tilde{h}(\tau) = h(s_0 \times \tau)$. Thus we have

$$\rho'(s_0, d\tau) = \tilde{h}(\tau)\rho(s_0, d\tau) .$$

Then applying the DLR equations for ρ' and ρ, we find that $\rho'(s, d\tau) = \tilde{h}(\tau)\rho(s, d\tau)$ for all $s \in \Omega_{\Lambda}$. This implies $\rho' = \tilde{h}\rho$, so $h = \tilde{h}$ ρ-almost everywhere. Thus $h \in \mathfrak{B}_{\rho}$. ∎

THEOREM IV.3.10 (Lanford-Ruelle, [18]). *The set* Γ^{Φ} *is a Choquet simplex.*

Proof. Let C be the cone through Γ^{Φ}. Given $\rho \in C$, we must show that $C_\rho = \{\rho' \in C: \rho' \leq \rho\}$ is a lattice. (Note that the partial order in C is the usual partial order on measures, i.e., if $\rho = \rho' + \rho''$ with $\rho, \rho' \in C$ and ρ'' any positive measure, then $\rho'' \in C$; this is because the DLR equations are linear.) Now the proof of Lemma IV.3.9 actually shows that $C_\rho = \{h\rho : h \in \mathfrak{B}_\rho, 0 \leq h \leq 1\}$. It is clear that the sup and inf of any two functions in $\mathfrak{B}_\rho$ are also in $\mathfrak{B}_\rho$, so that this is a lattice. ∎

Next we examine the quantum system, with Γ the set of KMS states (with a fixed strongly continuous one-parameter group of "time-translation" automorphisms α_t). Here, in contrast with the classical case, the fact that we are on a lattice will play no role in the theory (beyond Lemma IV.3.6, which showed that $\mathfrak{B}_\rho$ was the center of $\pi(\mathfrak{A})''$). For the remainder of this section, $\mathfrak{A}$ will be any C*-algebra with identity with a strongly continuous one-parameter group of automorphisms α_t. Our goal will be to show that a KMS state is extremal if and only if it is a *factor state*, i.e., the center $\mathfrak{B}_\rho = \pi(\mathfrak{A})' \cap \pi(\mathfrak{A})''$ of $\pi(\mathfrak{A})''$ consists only of scalar multiples of 1; and that Γ is a Choquet simplex.

Now the "non-commutative ergodic theory" we developed in Section IV.2 for lattice translations carries over (with slight changes in a few proofs) to time translations. Thus for a given time-invariant state ρ we have a one-parameter group of unitary operators $U(t)$ on the Hilbert space $\mathcal{H}$ of the GNS construction, with

$$U(t)w = w \quad \text{and} \quad \pi(\alpha_t A) = U(t)\pi(A)U(t)^{-1} .$$

Moreover, the strong continuity of α_t is easily seen to imply strong continuity of $U(t)$. The projection P on time-invariant vectors of $\mathcal{H}$ is given by

$$P = \text{s-lim}\,(2N)^{-1} \int_{-N}^{N} U(t)\,dt .$$

The algebra $\mathfrak{R}$ of Section IV.2 is replaced by $\mathfrak{R}_1 = (\pi(\mathfrak{A}) \cup U(\mathbf{R}))''$. We also consider the algebra $\mathfrak{R}_2 = (\pi(\mathfrak{A})' \cup U(\mathbf{R}))''$. In analogy with the first half of Lemma IV.2.6, we have both

$$(11) \qquad P\mathfrak{R}_1P = (P\pi(\mathfrak{A})P)'' \text{ and } P\mathfrak{R}_2P = (P\pi(\mathfrak{A})'P)''$$

the latter being due to the fact that if $C \in \pi(\mathfrak{A})'$, then also $U(t)CU(t)^{-1} \in \pi(\mathfrak{A})'$. However, here we do *not* in general have $(P\pi(\mathfrak{A})P)''$ abelian, and extremal time-invariant states need not be ergodic for the time evolution. This is most easily seen in finite systems: let $\mathfrak{A}$ be the algebra of linear operators on a finite-dimensional Hilbert space, and take the time evolution $\alpha_t(A) = e^{itH}Ae^{-itH}$ for a (self-adjoint) Hamiltonian H. The extremal time-invariant states are the vector states for the eigenvectors of H. Note that for such states we are already in the situation described by the GNS construction; for an eigenvector w with $Hw = Ew$ we have $U(t) = e^{it(H-E1)}$ and P is the projection on the eigenspace for E. Thus if E is degenerate, $P\mathfrak{A}P$ is non-abelian and the state is non-ergodic. On the other hand, for a "weakly asymptotically abelian" system factor states are ergodic (see [25], Exercise 6.E(c)).

LEMMA IV.3.11. *For any time-invariant state* ρ,

$$P\mathfrak{R}_1P = (P\pi(\mathfrak{A})'P)' = P\mathfrak{R}_2'P .$$

Proof. By (11) we have $P\mathfrak{R}_1P = (P\pi(\mathfrak{A})P)''$, while (using Lemma IV.2.5) $(P\pi(\mathfrak{A})'P)' = (P\mathfrak{R}_2P)' = P\mathfrak{R}_2'P$. To show $P\pi(\mathfrak{A})P$ commutes with $P\pi(\mathfrak{A})'P$, we note that for $A \in \mathfrak{A}$, $B \in \pi(\mathfrak{A})'$

$$P\pi(A)PBP = \operatorname*{s-lim}_{N\to\infty} \frac{1}{2N}\int_{-N}^{N} P\pi(\alpha_tA)BP\,dt = \operatorname*{s-lim}_{N\to\infty} \frac{1}{2N}\int_{-N}^{N} PB\pi(\alpha_tA)P\,dt$$

$$= PBP\pi(A)P .$$

Thus $(P\pi(\mathfrak{A})P)'' \subset (P\pi(\mathfrak{A})'P)'$. Finally, since $\pi(\mathfrak{A}) \subset \mathfrak{R}_1$ we have $(P\pi(\mathfrak{A})'P)' \subset (P\mathfrak{R}_1'P)' = P\mathfrak{R}_1P$. ■

THEOREM IV.3.12 (see [3], [26]). *Let* ρ *be a KMS state. Then there is a one-to-one affine correspondence between the set* C_ρ *of positive linear functionals* $\phi \leq \rho$ *satisfying the KMS conditions, and the set* $\{T \in \mathfrak{B}_\rho : 0 \leq T \leq 1\}$, *given by*

$$\phi(A) = \langle w, T\pi(A)w \rangle . \tag{12}$$

Thus Γ *is a Choquet simplex, and* ρ *is extremal in* Γ *if and only if* $\mathfrak{B}_\rho = \mathbf{C}1$.

Proof. If $T \in \mathfrak{B}_\rho$ with $0 \leq T \leq 1$, it is clear that $0 \leq \phi \leq \rho$. Moreover, if T is the strong limit of a net $\pi(C_\alpha)$ with $C_\alpha \in \mathfrak{A}$, then for any $A \in \mathfrak{A}$ and $B \in \tilde{\mathfrak{A}}$ we have

$$\phi(AB) = \lim_\alpha \rho(C_\alpha AB) = \lim_\alpha \rho(\alpha_{-i}(B)C_\alpha A) = \phi(\alpha_{-i}(B)A) .$$

Thus ϕ satisfies the KMS conditions (as in formula (11) of Section III.3).

Conversely, suppose ϕ satisfies the KMS conditions and $0 \leq \phi \leq \rho$. Then, as in the proof of Lemma IV.2.4 we obtain $T \in \pi(\mathfrak{A})'$ with $0 \leq T \leq 1$ such that (12) holds. Being KMS, ϕ is time-invariant, so T commutes with all the $U(t)$, and thus also with P. Note that if $A \in \mathfrak{A}$ and $B \in \tilde{\mathfrak{A}}$, $\langle w, T\pi(A)P\pi(\alpha_t B)w\rangle$ is constant (as a function of $t \in \mathbf{R}$). Thus the entire function $\langle w, T\pi(A)P\pi(\alpha_z B)w\rangle$ is also constant, and

$$\langle w, T\pi(A)P\pi(B)w\rangle = \langle w, T\pi(A)P\pi(\alpha_i B)w\rangle =$$

$$= \lim_{N\to\infty} \frac{1}{2N}\int_{-N}^{N} \langle w, T\pi(A)\,U(t)\,\pi(\alpha_i B)w\rangle\, dt$$

$$= \lim_{N\to\infty} \frac{1}{2N}\int_{-N}^{N} \phi(A\alpha_{t+i}B)\, dt = \lim \frac{1}{2N}\int_{-N}^{N} \phi(\alpha_t(B)A)\, dt = \langle w, T\pi(B)P\pi(A)w\rangle .$$

By continuity this is true for any $B \in \mathfrak{A}$. A similar calculation, using the KMS conditions for ρ, shows that $\langle w, \pi(A)P\pi(B)w\rangle = \langle w, \pi(B)P\pi(A)w\rangle$. Now suppose $C \in \mathfrak{R}_1'$ is self-adjoint, and $C_n \in \mathfrak{A}$ with $\pi(C_n)w \to Cw$. Then we obtain

$$\langle w, (\pi(C_n^*)-C)P(\pi(C_n)-C)w\rangle = \langle w, (\pi(C_n)-C)P(\pi(C_n^*)-C)w\rangle$$

so that $P\pi(C_n^*)w \to Cw$. Now for $A, B \in \mathfrak{A}$

$$\langle w, \pi(A)CT\pi(B)w\rangle = \langle Cw, TP\pi(AB)w\rangle = \lim_{n\to\infty} \langle w, T\pi(C_n^*)P\pi(AB)w\rangle$$

$$= \lim_{n\to\infty} \langle w, T\pi(AB)P\pi(C_n^*)w\rangle = \langle w, T\pi(AB)Cw\rangle = \langle w, \pi(A)TC\pi(B)w\rangle .$$

Thus $T \in \mathfrak{R}_1$. By Lemma IV.3.11, there is $S \in \mathfrak{R}_2'$ with $T\restriction_{P\mathcal{H}} = S\restriction_{P\mathcal{H}}$. Now the KMS conditions for ρ extend to the functional $A \mapsto \langle w, Aw\rangle$ on $\pi(\mathfrak{A})''$. Since in particular $S \in \pi(\mathfrak{A})''$, we have for $A \in \mathfrak{A}$, $B \in \tilde{\mathfrak{A}}$

$$\langle w, \pi(A)S\pi(B)w\rangle = \langle w, \pi(\alpha_{-i}(B)A)Sw\rangle = \phi(\alpha_{-i}(B)A) = \phi(AB)$$

$$= \langle w, \pi(A)T\pi(B)w\rangle .$$

Thus $T = S \in \mathfrak{B}_\rho$. ∎

IV.4. *Orthogonal decomposition.*

So far in this chapter we have only considered the weak-* topology on states. This has some physical justification (an experiment only measures a finite number of observables), as well as the mathematical advantages that arise from compactness. In certain situations, however, the norm topology is also useful. For example, we have Lemma II.1.1 relating this norm to the norm on the dual of the space $\mathcal{B}$ of interactions; a further application will be given in Theorem V.2.1. Our basic tool in this area is the "orthogonal" decomposition of a hermitian linear functional on a C^*-algebra into positive and negative parts. In the classical case, this is the Jordan decomposition of a signed measure into its positive and negative parts.

THEOREM IV.4.1 (see [13]). *Let* ϕ *be a hermitian linear functional on a* C^*-*algebra* $\mathfrak{A}$. *Then there is a unique decomposition* $\phi = \phi_+ - \phi_-$ *with* $\phi_\pm$ *positive linear functionals and* $\|\phi\| = \|\phi_+\| + \|\phi_-\|$.

Proof. We can consider ϕ as a functional on the real Banach space $\mathfrak{A}_h$ of self-adjoint elements of $\mathfrak{A}$ (as shown in the proof of Lemma II.1.1, this restriction has norm $\|\phi\|$). Let P be the compact convex set of positive linear functionals on $\mathfrak{A}_h$ with norm ≤ 1 (in the dual $\mathfrak{A}_h^*$ with the weak-* topology). Consider the polar

$$P^\circ = \{A \in \mathfrak{A}_h : |\rho(A)| \leq 1 \text{ for all } \rho \in P\}$$

of P (since we are using the weak-* topology on $\mathfrak{A}_h^*$, for which all continuous linear functionals are given by elements of $\mathfrak{A}_h$, we take P° in $\mathfrak{A}_h$). We claim that P° is the closed unit ball of $\mathfrak{A}_h$, i.e., that for any $A \in \mathfrak{A}_h$ there is a state ρ with $|\rho(A)| = \|A\|$. For this purpose, an identity element can be adjoined to $\mathfrak{A}$ if it is not already present. Choosing $\pm\|A\|$ in the spectrum $\sigma(A)$ of A, we can use the Hahn-Banach Theorem to separate the linear span of $(1,1)$ and $(A, \pm\|A\|)$ from the open convex set $\{(B,t): \sigma(B) \subset (t,\infty)\}$ in $\mathfrak{A}_h \oplus \mathbf{R}$. The hyperplane obtained is $\{(B, \rho(B))\}$ for a suitable state ρ, proving the claim. Therefore the "bipolar" $P^{\circ\circ}$ is the closed unit ball of $\mathfrak{A}_h^*$. But the "Bipolar Theorem" (see [33], p. 137) says that $P^{\circ\circ}$ is the balanced weak-* closed convex hull of P. Since P is convex and compact, the set

$$\{s\rho - (1-s)\rho' : \rho, \rho' \in P,\ 0 \leq s \leq 1\}$$

is already balanced, weak-* compact and convex, and so this is $P^{\circ\circ}$. Thus if $\|\phi\| = 1$ we have $\phi = s\rho - (1-s)\rho'$ with $\rho, \rho' \in P$, so that $\|s\rho\| + \|(1-s)\rho'\| \leq 1$. This proves existence.

Now suppose we have two decompositions $\phi = \phi_+ - \phi_- = \phi'_+ - \phi'_-$ satisfying the hypotheses. Consider the GNS representation for a state that dominates ϕ_+, ϕ_-, ϕ'_+ and ϕ'_-. These, and ϕ, have unique weakly continuous extensions to $\pi(\mathfrak{A})''$ with the same norms. Since the closed unit ball of $\pi(\mathfrak{A})''_h$ is compact in the weak operator topology, there is $B \in \pi(\mathfrak{A})''_h$ with $\|B\| = 1$ and $\phi(B) = \|\phi\|$. Consider the positive and negative parts $B^+ = P^+B$, $B^- = (P^+ - 1)B$ of B, where P^+ is a projection in $\pi(\mathfrak{A})''$ which commutes with B. We have

$$\phi_+(B^+) + \phi_-(B^-) - \phi_+(B^-) - \phi_-(B^+) = \phi(B) = \|\phi_+\| + \|\phi_-\|$$

with

$$\phi_+(B^+) \le \|\phi_+\|, \qquad \phi_-(B^-) \le \|\phi_-\|$$

and

$$-\phi_+(B^-) - \phi_-(B^+) \le 0$$

so that $\phi_+(B^+) = \|\phi_+\|$ and $\phi_-(B^-) = \|\phi_-\|$. Since $B^+ \le P^+ \le 1$ we have $\phi_+(P^+) = \phi_+(1) = \|\phi_+\|$. Thus for any $A \in \pi(\mathfrak{A})''$

$$|\phi_+((1-P^+)A)| \le |\phi_+(1-P^+)|^{\frac{1}{2}}\, |\phi_+(A^*A)|^{\frac{1}{2}} = 0$$

(using the Cauchy-Schwartz Inequality), so that $\phi_+(A) = \phi_+(P^+A)$. Similarly, using $B^- \le 1 - P^+ \le 1$ we find that $\phi_-(A) = \phi_-((1-P^+)A)$ and $\phi_-(P^+A) = 0$. Thus we obtain

$$\phi_+(A) = \phi(P^+A) \text{ and } \phi_-(A) = \phi((P^+ - 1)A) .$$

The same is true for ϕ'_+ and ϕ'_-, so $\phi'_+ = \phi_+$ and $\phi'_- = \phi_-$. ■

Note that the uniqueness of the decomposition implies that ϕ_+ and ϕ_- are invariant under any *-automorphisms of the algebra under which ϕ is invariant.

We can use this theorem to obtain Lipschitz estimates on affine maps. Suppose K is a convex set of states with the property that for any $\rho, \rho' \in K$, the positive and negative parts of $\rho - \rho'$ are in the cone through K (i.e., $\rho - \rho' = \frac{1}{2}\|\rho-\rho'\|(\rho_+ - \rho_-)$ with $\rho_+, \rho_- \in K$). This is true of the set E^I of invariant states under any group of *-automorphisms; we shall see below that it is also true for the set of KMS or DLR states for an interaction of our lattice systems. Suppose f is an affine function from K into any normed linear space, with $\|f(\rho)-f(\rho')\| \le N$ for all $\rho, \rho' \in K$. Then for ρ_+, ρ_- as above, we obtain

$$\|f(\rho)-f(\rho')\| = \frac{1}{2}\|\rho-\rho'\| \, \|f(\rho_+)-f(\rho_-)\| \le \frac{N}{2}\|\rho-\rho'\| .$$

In particular, if K is a (weak-*) compact and metrizable Choquet simplex, consider the affine map $\rho \mapsto \mu_\rho$ where μ_ρ is the unique probability measure on $\mathcal{E}(K)$ with barycenter ρ. Here we have the above estimates with $N = 2$. On the other hand, it is clear that $\|\rho-\rho'\| \le \|\mu_\rho-\mu_{\rho'}\|$.

COROLLARY IV.4.2. *The ergodic decomposition of Theorem IV.3.3 is an isometry of* E^I *onto the space* $\mathfrak{M}_1(\mathcal{E}(E^I))$ *of probability measures on* $\mathcal{E}(E^I)$.

The second case we consider is the mean entropy. In the weak-* topology this is upper semicontinuous, but examples can easily be constructed to show that it is not continuous. We consider the classical discrete-spin system (with μ_0 normalized counting measure) or the quantum system. Thus $0 \ge s(\rho) \ge -\ln n$ for all $\rho \in E^I$, where n is $|\Omega_0|$ or $\dim \mathcal{H}_0$.

COROLLARY IV.4.3. *In the classical discrete-spin or quantum system with* n *as above,* $|s(\rho)-s(\rho')| \le \frac{\ln n}{2}\|\rho-\rho'\|$ *for all* $\rho, \rho' \in E^I$.

In the context of a Choquet simplex of states, there is a more explicit decomposition into positive and negative parts. If the cone C is a lattice in its own order, for any $\phi \in C-C$ we have $\phi = \phi_1 - \phi_2$ with $\phi_1 = \phi \vee 0 \in C$ and $\phi_2 = (-\phi) \vee 0 \in C$. Clearly if $\phi = \phi_1' - \phi_2'$ with $\phi_1', \phi_2' \in C$ we have $\phi_1' \ge \phi_1$ and $\phi_2' \ge \phi_2$. Thus (taking C as the cone through a Choquet simplex of states) if ϕ_+ and ϕ_- are also in C, they must be ϕ_1 and ϕ_2 respectively. On the other hand, it may not be easy to compute $\|\phi_1\| + \|\phi_2\| = \|\phi\|$ (e.g., for E^I in the quantum lattice system) without using Theorem IV.4.1.

Suppose F is a *face* of the simplex K (i.e., F is convex, and if $\rho, \rho' \in K$ with $\frac{\rho+\rho'}{2} \in F$, then $\rho, \rho' \in F$). Let C_F be the cone through F. It is easy to see that the inf and sup of any two elements of $C_F - C_F$ (for the order defined by C) are in $C_F - C_F$ as well. Considering in particular the set Δ_Φ of invariant equilibrium states for an interaction $\Phi \in \mathcal{B}$ (this is a face of E^I since it is the set on which the affine function $\rho \mapsto s(\rho) - \rho(A_\Phi)$ attains its maximum), we have

LEMMA IV.4.4. *If ρ and ρ' are positive multiples of invariant equilibrium states for the interaction $\Phi \in \mathcal{B}$, then so are the positive and negative parts of $\rho - \rho'$.*

Next we consider the DLR equations. These equations can be considered for any measure on Ω, not just a state.

LEMMA IV.4.5. *Let ϕ be any complex measure on Ω which satisfies the DLR equations for an interaction $\Phi \in \tilde{\mathcal{B}}$. Then the positive and negative real and imaginary parts of ϕ also satisfy the DLR equations for Φ.*

Proof. Since the DLR equations (as given in equation (3) of Section III.1) are linear equations in the measures $\phi(s,d\tau)$ with real coefficients, the real and imaginary parts of ϕ will satisfy the DLR equations. Thus we can assume ϕ is real. Let Λ be a fixed finite region. We have $\phi(ds \times d\tau) = \phi(s,d\tau)\mu_0(ds)$ (interpreted in the sense of equation (2) of Section III.1), with

$$\phi(s,d\tau) = e^{-W(s,s' \times \tau)} \phi(s',d\tau) .$$

Letting $\phi(s,d\tau)_+$ denote the positive part of the measure $\phi(s,d\tau)$, we have the same relation between $\phi(s,d\tau)_+$ and $\phi(s',d\tau)_+$. Now is the discrete-spin case it is clear that the positive part of ϕ is $\phi^+(ds \times d\tau) = \phi(s,d\tau)_+ \mu_0(ds)$. In the general case, a bit of caution might be needed. First, ϕ^+ is a well-defined measure since for any fixed $s_0 \in \Omega_\Lambda$

$$\int_{\Omega_{\Lambda^c}} f(s\times\tau)\phi(s,d\tau)_+ = \int_{\Omega_{\Lambda^c}} f(s\times\tau)e^{-W(s,s_0\times\tau)}\phi(s_0,d\tau)_+$$

is a continuous function of s for each $f \in C(\Omega)$. Next, to show that ϕ^+ is the positive part $\phi_+ = \phi \vee 0$ of ϕ: clearly $\phi^+ \geq \phi$ and $\phi^+ \geq 0$, and for any $f \in C(\Omega)$ with $f \geq 0$ we have

$$\phi^+(f) = \int_{\Omega_{\Lambda^c}} \int_{\Omega_\Lambda} f(s\times\tau)e^{-W(s,s_0\times\tau)}\mu_0(ds)\phi(s_0,d\tau)_+$$

$$= \sup_{\substack{g\in C(\Omega_{\Lambda^c})\\ 0\leq g\leq 1}} \int_{\Omega_{\Lambda^c}} g(\tau)\int_{\Omega_\Lambda} f(s\times\tau)e^{-W(s,s_0\times\tau)}\mu_0(ds)\phi(s_0,d\tau)$$

$$= \sup_{\substack{g\in C(\Omega_{\Lambda^c})\\ 0\leq g\leq 1}} \phi(gf) \leq \phi_+(f)\,.$$

Thus $\phi_+ = \phi^+$, which satisfies the DLR equations, and so does ϕ_-, which is $\phi_+ - \phi$. ∎

For the KMS conditions, we do not have such a strong result, since we need the added condition that ϕ is dominated by a KMS state ρ in the sense that for some N,

$$|\phi(A^*B)| \leq N\rho(A^*A)^{\frac{1}{2}}\rho(B^*B)^{\frac{1}{2}} \quad \text{for all } A,B \in \mathfrak{A}\,. \tag{13}$$

This condition will be satisfied if ϕ is a difference of two positive KMS functionals ρ_1 and ρ_2, where ρ can be taken as $\|\rho_1+\rho_2\|^{-1}(\rho_1+\rho_2)$.

LEMMA IV.4.6. *Suppose the linear functional ϕ on $\mathfrak{A}$ satisfies the KMS conditions, and also* (13) *for some* N *and some KMS state* ρ. *Then the* *positive and negative real and imaginary parts of* ϕ *also satisfy the KMS conditions.*

Proof. The linear functional $\phi^*(A) \equiv \overline{\phi(A^*)}$ is easily seen to satisfy (13) and the KMS conditions: e.g., if we take functions F_{AB} for ϕ as in equation (12) of Section III.3, the corresponding functions for ϕ^* will be $G_{AB}(z) = \overline{F_{A^*B^*}(\bar{z}+i)}$. Thus the real and imaginary parts of ϕ also satisfy the same conditions, so we can assume ϕ is hermitian. Taking the GNS construction for the state ρ, the estimate (13) means that the sesquilinear form

$$Q(\pi(A)w, \pi(B)w) = \phi(A^*B)$$

is bounded, and so ϕ is given by $\phi(A) = <w, T\pi(A)w>$ for some $T \epsilon \pi(\mathfrak{A})'$ self-adjoint. As in the proof of Theorem IV.3.12, we obtain $T \epsilon \mathfrak{B}_\rho = \pi(\mathfrak{A})' \cap \pi(\mathfrak{A})''$. The positive and negative parts T_+ and T_- of T are also in $\mathfrak{B}_\rho$, so that the positive functionals

$$\phi_1(A) = <w, T_+ \pi(A)w> \qquad \phi_2(A) = <w, T_- \pi(A)w>$$

on $\mathfrak{A}$ are KMS. To complete the proof, we must show that these are in fact the positive and negative parts of ϕ, by showing ϕ has norm at least $\|\phi_1\| + \|\phi_2\| = \phi_1(1) + \phi_2(1) = <w, |T|w>$. But $|T| = TU$ with $U \epsilon \mathfrak{B}_\rho$ and $\|U\| = 1$. By the Kaplansky Density Theorem and Lemma IV.3.8, U is the strong limit of a net $\pi(A_\alpha)$ with $\|A_\alpha\| \leq 1$, so

$$<w, |T|w> = <w, TUw> = \lim_\alpha \phi(A_\alpha) \leq \|\phi\|. \qquad \blacksquare$$

COROLLARY IV.4.7. *The map* $\rho \rightarrow \mu_\rho$ *from the set* Γ *of DLR states for an interaction of the classical system, or KMS states for a time evolution* a_t, *to the space* $\mathfrak{M}_1(\mathcal{E}(\Gamma))$ *of probability measures on* $\mathcal{E}(\Gamma)$, *is an isometry.*

Proof. See the remarks above Corollary IV.4.2. $\qquad \blacksquare$

V. APPROXIMATION BY TANGENT FUNCTIONALS: EXISTENCE OF PHASE TRANSITIONS

V.1. *The theorem of Bishop and Phelps*

A standard type of problem in statistical mechanics is to describe the invariant equilibrium states for a given interaction. In this chapter we consider the reverse situation: given an invariant state ρ, we look for an interaction having an invariant equilibrium state $\tilde{\rho}$ which bears some resemblance to ρ. In particular, if ρ exhibits some type of long-range order, we would like $\tilde{\rho}$ to share this property. This approach leads to some very general existence results for phase transitions, showing that a given type of phase transition occurs for some member of a certain class of interactions. In Section V.2, this class will be the whole Banach space $\mathcal{B}$, for which very pathological behavior will be found. In Section V.3 we consider much smaller classes of interactions, such as ferromagnetic pair interactions, and find a more "reasonable" type of phase transition.

As we have seen, invariant states determine linear functionals on the space of interactions. For states of finite mean entropy, these functionals are P-bounded (Sections II.1 and II.3). We want to approximate such a functional by a tangent functional to the pressure, and thus approximate an invariant state by an invariant equilibrium state for some interaction. Our basic tool is a version of a theorem of Bishop and Phelps [4].

THEOREM V.1.1. *Let* P *be a lower semicontinuous convex function on a real Banach space* $\mathcal{X}$ *(we can allow* P *to take the value* $+\infty$ *but assume* P *is not identically* $+\infty$*). Then tangent functionals to* P *are norm-dense in the* P*-bounded linear functionals on* $\mathcal{X}$. *In fact if* $a_0 \in \mathcal{X}^*$ *is* P*-bounded and* $\Phi_0 \in \mathcal{X}$ *with* $P(\Phi_0) < \infty$, *then for any* $\varepsilon > 0$ *there is*

$\tilde{a} \in \mathfrak{X}^*$ *which is tangent to* P *at some* $\tilde{\Phi} \in \mathfrak{X}$ *with* $\|a_0 - \tilde{a}\| \leq \varepsilon$ *and* $\|\Phi_0 - \tilde{\Phi}\| \leq \varepsilon^{-1}(P(\Phi_0) - a_0(\Phi_0) - S(a_0))$ *where*

$$S(a_0) = \inf\{P(\Psi) - a_0(\Psi) : \Psi \in \mathfrak{X}\} .$$

(Note that in statistical mechanics, $S(a_0)$ will be the mean entropy of the state corresponding to a_0.)

Proof. Instead of P we will consider the function $\Psi \mapsto P(\Psi) - a_0(\Psi) - S(a_0)$, so that the theorem is reduced to the case where $a_0 = 0$ and $S(a_0) = 0$. Now we must find $\tilde{a}$ tangent to our new function P at $\tilde{\Phi}$ with $\|\tilde{a}\| \leq \varepsilon$ and $\|\Phi_0 - \tilde{\Phi}\| \leq \varepsilon^{-1} P(\Phi_0)$. See Figure 3.

For each $\Phi \in \mathfrak{X}$ we define

$$C(\Phi) = \{\Psi \in \mathfrak{X} : P(\Psi) \leq P(\Phi) - \varepsilon\|\Phi - \Psi\|\} .$$

Thus $\Psi \in C(\Phi)$ if and only if the corresponding point $(\Psi, P(\Psi))$ on the graph of P lies in the closed downward-pointing cone with vertex $(\Phi, P(\Phi))$ and slope ε. We will find $\tilde{\Phi}$ so that $C(\tilde{\Phi}) = \{\tilde{\Phi}\}$, i.e., the cone with vertex $(\tilde{\Phi}, P(\tilde{\Phi}))$ will touch the graph of P only at the vertex.

Clearly $C(\Phi)$ is closed, and if $\Psi \in C(\Phi)$ then $C(\Psi) \subset C(\Phi)$ (the cones are nested). We construct a sequence $\Phi_0, \Phi_1, \cdots,$ by choosing $\Phi_{n+1} \in C(\Phi_n)$ with

$$P(\Phi_{n+1}) < 2^{-n}\varepsilon + \inf\{P(\Psi) : \Psi \in C(\Phi_n)\} .$$

Then if $\Psi \in C(\Phi_{n+1})$, the above condition implies

$$P(\Phi_{n+1}) - 2^{-n}\varepsilon < P(\Psi) \leq P(\Phi_{n+1}) - \varepsilon\|\Psi - \Phi_{n+1}\|$$

so that $\|\Psi - \Phi_{n+1}\| < 2^{-n}$. Thus the Φ_n form a Cauchy sequence, and have a limit $\tilde{\Phi} \in \bigcap_n C(\Phi_n)$. If $\Psi \in C(\tilde{\Phi}) \subset \bigcap_n C(\Phi_n)$, then $\|\Psi - \Phi_{n+1}\| < 2^{-n}$ for all n, so that $\Psi = \tilde{\Phi}$. Thus $C(\tilde{\Phi}) = \{\tilde{\Phi}\}$ as desired.

Now the open cone $\{(\Psi, y) \in \mathfrak{X} \oplus \mathbf{R} : y < P(\tilde{\Phi}) - \varepsilon\|\Psi - \tilde{\Phi}\|\}$ and the "super-graph" $\{(\Psi, y) : y \geq P(\Psi)\}$ of P are convex and disjoint, so by the Hahn-

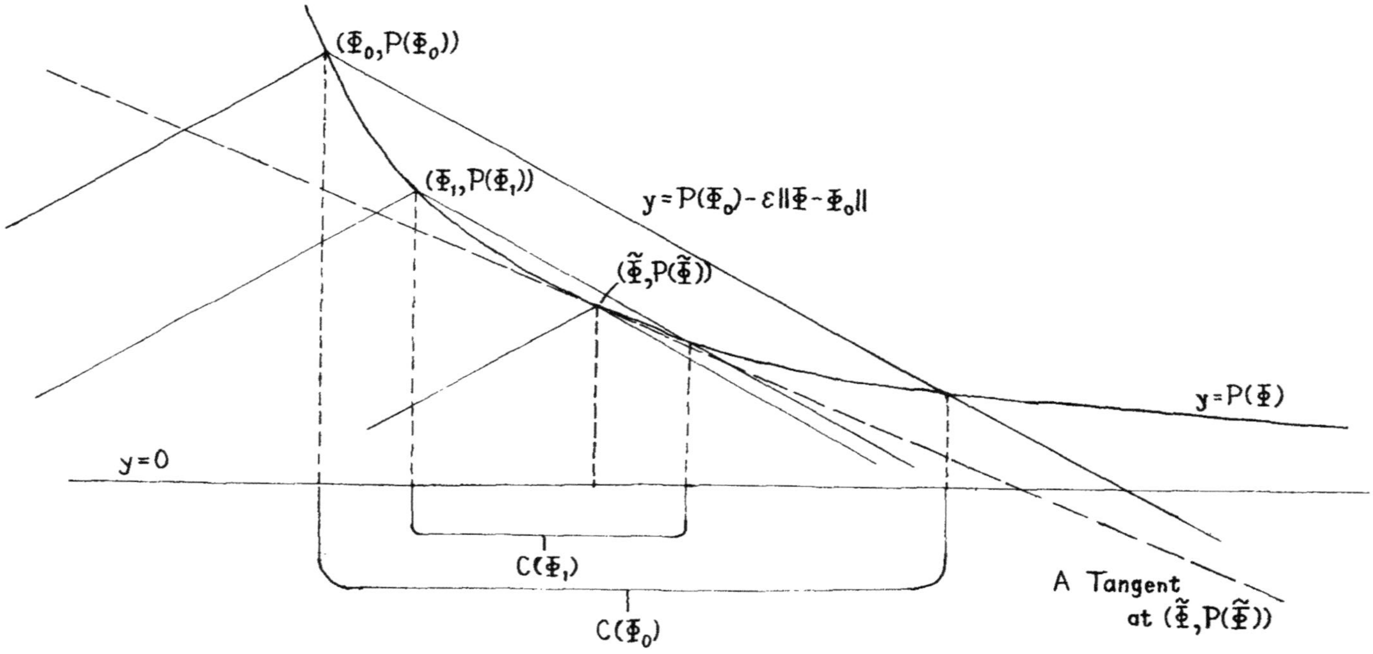

Figure 3

Banach Theorem they can be separated by a nonzero continuous linear functional on $\mathfrak{X} \oplus \mathrm{R}$. This functional can be written as $(\Psi, y) \mapsto y - \tilde{\alpha}(\Psi)$, so that for some $t \in \mathrm{R}$

$$P(\tilde{\Phi}) - \varepsilon \|\Psi - \tilde{\Phi}\| - \tilde{\alpha}(\Psi) \leq t \leq P(\Psi) - \tilde{\alpha}(\Psi) \quad \text{for all } \Psi \in \mathfrak{X} .$$

Taking in particular $\Psi = \tilde{\Phi}$ we have $t = P(\tilde{\Phi}) - \tilde{\alpha}(\tilde{\Phi})$, so the above inequalities become

$$\tilde{\alpha}(\tilde{\Phi}) - \tilde{\alpha}(\Psi) \leq \varepsilon \|\Psi - \tilde{\Phi}\|$$

$$P(\tilde{\Phi}) - \tilde{\alpha}(\tilde{\Phi}) + \tilde{\alpha}(\Psi) \leq P(\Psi) .$$

The first says $\|\tilde{\alpha}\| \leq \varepsilon$ and the second says $\tilde{\alpha}$ is tangent to P at $\tilde{\Phi}$. Finally, since $\tilde{\Phi} \in C(\Phi_0)$ we have

$$\|\Phi_0 - \tilde{\Phi}\| \leq \varepsilon^{-1} (P(\Phi_0) - P(\tilde{\Phi})) \leq \varepsilon^{-1} P(\Phi_0) . \qquad \blacksquare$$

In most applications, we will not need such strong control over $\tilde{\alpha}$ as is given by this theorem, but we will want much more control over $\tilde{\Phi}$. Thus we may wish only to approximate the functional α_0 on some closed subspace of $\mathfrak{X}$, or we may only need a one-sided estimate $\tilde{\alpha}(\Psi) \geq \alpha_0(\Psi) - \varepsilon \|\Psi\|$ for Ψ in some closed cone. Meanwhile we would like $\tilde{\Phi}$ to lie in some closed subspace or cone (e.g., the cone of "ferromagnetic" pair interactions). This can be accomplished by a slight modification of the proof, as in the following corollary. Note that here we must assume P is continuous, not just lower semicontinuous, since the cone we separate from the graph of P will no longer be open.

COROLLARY V.1.2. *Let $\mathfrak{F}$ be a closed convex cone in $\mathfrak{X}$ (with vertex 0), and suppose* P *is continuous. Then given $\Phi_0 \in \mathfrak{X}$, $\alpha_0 \in \mathfrak{X}^*$ P-bounded, and $\varepsilon > 0$, there is $\tilde{\alpha} \in \mathfrak{X}^*$ which is tangent to* P *at some $\tilde{\Phi} \in \Phi_0 + \mathfrak{F}$, with $\|\Phi_0 - \tilde{\Phi}\| \leq \varepsilon^{-1}(P(\Phi_0) - \alpha_0(\Phi_0) - S(\alpha_0))$ and*

$$\tilde{\alpha}(\Psi) \geq \alpha_0(\Psi) - \varepsilon \|\Psi\| \quad \text{for } \Psi \in \mathfrak{F} . \tag{1}$$

(Note that here we take $S(\alpha_0)$ as in Theorem V.1.1. Actually we could take $\inf\{P(\Psi)-\alpha_0(\Psi): \Psi \in \Phi_0 + \mathfrak{F}\}$ instead, but in the applications this does not improve our estimates.)

Proof. We modify the proof of Theorem V.1.1 by considering, instead of the entire downward-pointing cone of slope ε, only that part of it with azimuth in the directions given by $\mathfrak{F}$. Thus to construct the sequence $\Phi_0, \Phi_1, \cdots$, we use $C(\Phi) \cap (\Phi + \mathfrak{F})$ instead of $C(\Phi)$. The limit $\tilde{\Phi}$ then has the property $C(\tilde{\Phi}) \cap (\tilde{\Phi} + \mathfrak{F}) = \{\tilde{\Phi}\}$. Thus the open supergraph $\{(\Psi, y) \in \mathfrak{X} \oplus R: y > P(\Psi)\}$ and the new cone

$$\{(\Psi, y): \Psi \in \tilde{\Phi} + \mathfrak{F},\ y \leq P(\tilde{\Phi}) - \varepsilon \|\Psi - \tilde{\Phi}\|\}$$

are disjoint and can be separated by a hyperplane $y - \tilde{\alpha}(\Psi) = t$. Since $t = P(\tilde{\Phi}) - \tilde{\alpha}(\tilde{\Phi})$, we obtain the inequalities

$$\tilde{\alpha}(\Psi) - \tilde{\alpha}(\tilde{\Phi}) \geq -\varepsilon \|\Psi - \tilde{\Phi}\| \quad \text{for } \Psi \in \tilde{\Phi} + \mathfrak{F}$$

$$P(\tilde{\Phi}) - \tilde{\alpha}(\tilde{\Phi}) + \tilde{\alpha}(\Psi) \leq P(\Psi) \quad \text{for all } \Psi \in \mathfrak{X}.$$ ■

In particular, we can take the cone $\mathfrak{F}$ to be a closed linear subspace of $\mathfrak{X}$. The estimate (1) then means $\|(\tilde{\alpha} - \alpha_0){\restriction}_{\mathfrak{F}}\| \leq \varepsilon$.

V.2. *"Anti-phase transitions" in* $\mathcal{B}$

We now apply Theorem V.1.1 to statistical mechanics. Here we take $\mathfrak{X}$ as the space $\mathcal{B}$ of interactions for either the classical or quantum system, and P the pressure. Starting with an invariant state ρ_0 with $s(\rho_0) > -\infty$, we take the corresponding P-bounded functional α_0. The tangent functional $\tilde{\alpha}$ of Theorem V.1.1 corresponds in turn to an invariant equilibrium state $\tilde{\rho}$ for $\tilde{\Phi} \in \mathcal{B}$. By Lemma II.1.1, we have $\|\tilde{\rho} - \rho_0\| = \|\tilde{\alpha} - \alpha_0\|$. Thus in this case Theorem V.1.1 asserts the following:

COROLLARY V.2.1. *For any* $\Phi_0 \in \mathcal{B}$, $\rho_0 \in E^I$ *with* $s(\rho_0) > -\infty$, *and* $\varepsilon > 0$, *there is an invariant equilibrium state* $\tilde{\rho}$ *for some* $\tilde{\Phi} \in \mathcal{B}$ *with* $\|\rho_0 - \tilde{\rho}\| \leq \varepsilon$ *and* $|||\Phi_0 - \tilde{\Phi}||| \leq \varepsilon^{-1}(P(\Phi_0) + \rho_0(A_{\Phi_0}) - s(\rho_0))$.

This result gives us very strong control over $\tilde{\rho}$. Just how strong this control is, can best be seen using Corollary IV.4.2, which shows that if μ_0 and $\tilde{\mu}$ are the unique probability measures with barycenters ρ_0 and $\tilde{\rho}$ respectively, then $\|\mu_0-\tilde{\mu}\| = \|\rho_0-\tilde{\rho}\|$.

THEOREM V.2.2. a) *If* $\rho_1,\cdots,\rho_n$ *are ergodic states with* $s(\rho_j) > -\infty$, *there is some interaction* $\Phi \in \mathcal{B}$ *for which these are all invariant equilibrium states.*

b) *If* μ *is a nonatomic probability measure concentrated on the ergodic states with finite mean entropy, there is some* $\Phi \in \mathcal{B}$ *which has uncountably many ergodic equilibrium states in the support of* μ.

c) *There is a dense set of interactions in* $\mathcal{B}$ *with uncountably many ergodic equilibrium states.*

Proof. Suppose $\rho_0 \in E^I$ with $s(\rho_0) > -\infty$, and μ_0 is the corresponding probability measure on $\mathcal{E}(E^I)$. Let S be a set of ergodic states with $\mu_0(S) > \varepsilon > 0$. Then if $\tilde{\rho}$ is an invariant equilibrium state for some $\tilde{\Phi} \in \mathcal{B}$ with $\|\rho_0-\tilde{\rho}\| \leq \varepsilon$, we will have $\tilde{\mu}(S) > 0$ where $\tilde{\mu}$ is the probability measure on $\mathcal{E}(E^I)$ with barycenter $\tilde{\rho}$. But by Theorem IV.3.3, $\tilde{\mu}$ is concentrated on invariant equilibrium states for $\tilde{\Phi}$, so S contains some of these. In (a) we let $\rho_0 = (\rho_1+\cdots+\rho_n)/n$ and take $\varepsilon < 1/n$. Then for $\tilde{\Phi}$ and $\tilde{\rho}$ as above, $\tilde{\mu}\{\rho_j\} > 0$ for each j, so all the ρ_j are invariant equilibrium states for $\tilde{\Phi}$. In (b) we take ρ_0 as the barycenter of μ, and $\varepsilon < 2$; this implies that there are uncountably many ergodic states in the intersection of the supports of μ and $\tilde{\mu}$, all of which are invariant equilibrium states for $\tilde{\Phi}$. This method, together with the estimate on $|||\Phi_0-\tilde{\Phi}|||$, will also give us (c) once we have proven the following lemma:

LEMMA V.2.3. *For any* $\Phi_0 \in \mathcal{B}$ *and* $\delta > 0$ *there is a nonatomic probability measure* μ *concentrated on ergodic states* ρ *with* $P(\Phi_0)+\rho(A_{\Phi_0})-s(\rho)\leq\delta$.

Proof. By Lemmas IV.3.1 and IV.3.2, ergodic states form a dense G_δ set in the compact metrizable space $\{\rho \in E^I: P(\Phi_0)+\rho(A_{\Phi_0})-s(\rho) \leq \delta\}$ (with the weak-* topology). Now it can be shown that any uncountable dense G_δ set in a compact metric space contains a homeomorphic image of the Cantor set. (The construction is a modification of the usual proof of the Baire Category Theorem. Suppose E is the intersection of a nested sequence of dense open sets U_n; by an inductive procedure we obtain a sequence of closed sets $K_n \subset U_n$, where K_n is the union of 2^n disjoint closed balls of radius less than 2^{-n}, the interior of each ball contains uncountably many points of E, and two balls of K_{n+1} are chosen in the interior of each ball of K_n. Then $\bigcap_n K_n$ is homeomorphic to the Cantor set.) Since the Cantor set (which is also homeomorphic to our space Ω for the spin-$\frac{1}{2}$ Ising model) has nonatomic probability measures, so does our G_δ. ∎

An alternate proof of this lemma uses the Gibbs Phase Rule, either in our version (Theorem VI.3.3) or the Gallavotti-Miracle version ([8], Theorem 3(ii)). This shows that there are $\Psi_0, \Psi_1 \in \mathcal{B}$ with $|||\Psi_i - \Phi_0||| \leq \delta/2$ such that $\Psi_t = t\Psi_1 + (1-t)\Psi_0$ has a unique invariant equilibrium state ρ_t for almost every $t \in [0,1]$. Moreover, we can take Ψ_0 and Ψ_1 in $\tilde{\mathcal{B}}$ (in the classical case) or $\mathcal{B}_r$ (quantum case), and can ensure that they are not physically equivalent, so that all the ρ_t are distinct. Since $|P(\Phi_0) - P(\Psi_t)| \leq |||\Phi_0 - \Psi_t|||$ and $P(\Psi_t) + \rho_t(A_{\Psi_t}) = s(\rho_t)$, we have

$$P(\Phi_0) + \rho_t(A_{\Phi_0}) - s(\rho_t) \leq 2|||\Phi_0 - \Psi_t||| \leq \delta .$$

For $F \in \mathfrak{A}_\Lambda$ self-adjoint, $\rho_t(F) = -\frac{d}{ds} P(\Psi_t + s\Psi_F)|_{s=0}$ for almost every t. As the pointwise (a.e.) limit of a sequence of continuous functions (difference quotients for this derivative), $\rho_t(F)$ is measurable as a function of t, and so is $t \mapsto \rho_t$ as an E^I-valued function. Lebesgue measure on $[0,1]$ thus induces a probability measure μ with the desired properties.

The Gibbs Phase Rule ([24]) shows that the interactions with non-unique invariant equilibrium states form a set of first category in $\mathcal{B}$ (or any of the other Banach spaces of interactions we have considered). Our

Theorem VI.3.3 will show that interactions with a large number of ergodic equilibrium states are in a sense quite rare: a "generic" k-dimensional subspace of $\mathfrak{B}$ will contain no interactions with more than $k+1$ ergodic equilibrium states. Nevertheless, by Theorem V.2.2 the set of interactions with uncountably many "pure phases" is dense in $\mathfrak{B}$. Now one of the goals of statistical mechanics is to show that large regions of the space of interactions are free of phase transitions; there is then some hope of understanding the phase transitions that do occur. Our result shows why this goal may be pursued only in smaller, better-behaved spaces such as $\mathfrak{B}_r$.

The most striking aspect of Theorem V.2.2 is not the occurrence of phase transitions, but of "anti-phase transitions" – the occurrence of a given ergodic state as an equilibrium state for many different interactions. This provides an interesting contrast to the results discussed in Section III.4. There it was shown that in $\tilde{\mathfrak{B}}$ (for the classical case) or $\mathfrak{B}_r$ (quantum case), the pressure is strictly convex except in the directions corresponding to physical equivalence, and two interactions with an invariant equilibrium state in common must be physically equivalent. These results fail in $\mathfrak{B}$ in the most spectacular manner imaginable. From any interaction in $\mathfrak{B}$ there are many directions in which the pressure has a linear segment. These directions are not related to physical equivalence; in fact they lead to interactions which share an invariant equilibrium state with the first, but also have additional ergodic equilibrium states which the first interaction lacks. Not only does an invariant equilibrium state not determine the interaction up to physical equivalence, it has no apparent influence on what other invariant equilibrium states the interaction may have. (There is one restriction on the set of ergodic equilibrium states for an interaction Φ: since the mean entropy is given on this set by $P(\Phi) + \rho(A_\Phi)$, it must be continuous on that set in the weak-* topology. Thus for some sequences of ergodic states, there is no interaction having the whole sequence as equilibrium states.) There are examples due to Fisher [7] of classical systems where two different interactions share an invariant equilibrium state; these have a physical interpretation in terms of the formation of a rigid crystal. On the other hand, there is no hope of physical interpretation for the bizarre behavior we have obtained, except as a pathology arising from the use of too large a space of interactions; it does not have any real connection to the phase transitions of more "reasonable" systems. The conclusion we must draw is that we should restrict our attention to the spaces $\tilde{\mathfrak{B}}$ and $\mathfrak{B}_r$, for which such pathologies are impossible.

V.3. *Existence of phase transitions*

We will now examine cases where the cone $\mathfrak{F}$ of Corollary V.1.2 contains a much smaller class of interactions; in particular, it will involve terms in at most a certain finite number of sites, so that the pathologies of Section V.2 are excluded. Instead, "reasonable" phase transitions will be obtained. Our results are valid for both classical and quantum systems; we will use the notation of the quantum system.

Suppose we wish to control the expectations of a set S of self-adjoint finite-range observables $B \in \mathfrak{A}_{\Lambda_B}$. To do this, we use the interactions $\Psi_B = \Psi_B^{\Lambda_B}$. Let $\mathfrak{F}$ be the closed convex cone generated by the Ψ_B (i.e., the closure of the set of linear combinations of the Ψ_B with positive coefficients). Then Corollary V.1.2 says the following:

COROLLARY V.3.1. *Let* $\Phi_0 \in \mathfrak{B}$, $\rho_0 \in E^I$ *with* $s(\rho_0) > -\infty$, *and* $\varepsilon > 0$ *be given. Then for* S *and* $\mathfrak{F}$ *as above, there is an invariant equilibrium state* $\tilde{\rho}$ *for some interaction* $\tilde{\Phi} \in \Phi_0 + \mathfrak{F}$, *with* $\tilde{\rho}(B) \leq \rho_0(B) + \varepsilon \|B\|$ *for* $B \in S$, *and* $|||\tilde{\Phi} - \Phi_0||| \leq \varepsilon^{-1}(P(\Phi_0) + \rho_0(A_{\Phi_0}) - s(\rho_0))$.

To obtain non-ergodic invariant equilibrium states or breaking of translation invariance from this type of information on $\tilde{\rho}$, we will use the results on "cluster properties" of states, which we presented in Chapter IV. Our first result will give a "ferromagnetic" type of phase transition, with terms linear and quadratic in translates of a given observable.

THEOREM V.3.2. *Let* $A \in \mathfrak{A}_X$ *be self-adjoint, and suppose there are invariant states of finite mean entropy that differ on* A. *Let* $\mathfrak{F}$ *be the cone of interactions* $\Psi \in \mathcal{B}$ *of the form*

$$\Psi(i+X) = h\,\tau_i A \qquad h \in R$$

$$\Psi((i+X)\cup(j+X)) = -J(i-j)(\tau_i A)(\tau_j A) \qquad J \geq 0 \tag{2}$$

$$\Psi(Y) = 0 \ \textit{for all other } Y.$$

In the quantum case, to make sure $(\tau_i A)(\tau_j A)$ *is self-adjoint, we require in addition* $J(i) = 0$ *unless* $(i+X)\cap X = \emptyset$. *Then for any* $\Phi_0 \in \mathcal{B}$ *there is* $\tilde{\Phi} \in \Phi_0 + \mathfrak{F}$ *which has at least two invariant equilibrium states that differ on* A.

Proof. Let $\rho_0 = (\rho_1 + \rho_2)/2$ for invariant states ρ_1, ρ_2 with $s(\rho_i) > -\infty$ and $\rho_1(A) \neq \rho_2(A)$. Then (with $c_n(A)$ defined as in equation (1) of Section IV.1)

$$\lim_{n\to\infty} \rho_0(c_n(A)^2) \geq \frac{1}{2}(\rho_1(A)^2 + \rho_2(A)^2) > \rho_0(A)^2 .$$

For any $\varepsilon > 0$, Corollary V.3.1 gives us an invariant equilibrium state $\tilde{\rho}$ for some $\tilde{\Phi} \in \Phi_0 + F$ with $|\tilde{\rho}(A) - \rho_0(A)| \leq \varepsilon \|A\|$ and, when $(i+X)\cap X = \emptyset$, $\tilde{\rho}(A\tau_i A) \geq \rho_0(A\tau_i A) - \varepsilon\|A\|^2$. Therefore

$$\lim_{n\to\infty} \tilde{\rho}(c_n(A)^2) \geq \lim_{n\to\infty} \rho_0(c_n(A)^2) - \varepsilon\|A\|^2$$

while

$$\tilde{\rho}(A)^2 \leq \rho_0(A)^2 + 2\varepsilon\|A\|^2 .$$

Thus for ε sufficiently small, $\tilde{\rho}$ is not ergodic. Moreover, if $\tilde{\mu}$ is the probability measure on $\mathcal{E}(E^I)$ with barycenter $\tilde{\rho}$, then

$$\int \rho'(A)^2 \tilde{\mu}(d\rho') = \lim_{n\to\infty} \int \rho'(c_n(A)^2)\tilde{\mu}(d\rho') = \lim_{n\to\infty} \tilde{\rho}(c_n(A)^2) > \tilde{\rho}(A)^2$$

so that $\rho'(A) \neq \tilde{\rho}(A)$ for some invariant equilibrium state ρ' in the support of $\tilde{\mu}$. ∎

Note that the condition for the interaction Ψ of (2) to be in $\mathfrak{B}$ is $\sum_i |J(i)| < \infty$. Since only terms in at most $2|X|$ sites are involved, Ψ is also in $\tilde{\mathfrak{B}}$ and all the $\mathfrak{B}_r$. However, we can not allow a norm which gives a higher weighting to terms $\Psi(Y)$ with large *diameter*; thus we can say nothing about the convergence of $\sum_i f(i)|J(i)|$ if $f(i) \to \infty$ as $|i| \to \infty$, without losing all information about correlations over large distances. This is because, if we tried to apply Corollary V.1.2 using such a norm, we would obtain the inequality $\tilde{\rho}(A\tau_i A) \geq \rho_0(A\tau_i A) - \varepsilon f(i) \|A\|^2$, which is useless when $\varepsilon f(i) \geq 2$. Thus our results are only for "long range" interactions; completely different methods are needed to deal with any stronger decrease of the interaction with distance.

To deal with breaking of translation invariance, we use the characterization of extremal DLR and KMS states as states with short range correlations (Lemma IV.3.9 and Theorem IV.3.12). We will say that *translation invariance is broken* for the interaction $\tilde{\Phi}$ if some ergodic equilibrium state for $\tilde{\Phi}$ fails to have short range correlations.

THEOREM V.3.3. *Let* $A \in \mathfrak{A}_X$ *and* $B \in \mathfrak{A}_Y$ *be self-adjoint, and suppose* $\lim_{|i| \to \infty} \rho_0(A\tau_i B)$ *fails to exist for some invariant state* ρ_0 *with* $s(\rho_0) > -\infty$. *Let* $\mathfrak{F}$ *be the closed linear span of the interactions* $\Psi_{A\tau_i B}$ *for* $X \cap (i+Y) = \emptyset$. *Then for any* $\Phi_0 \in \mathfrak{B}$ *there is* $\tilde{\Phi} \in \Phi_0 + \mathfrak{F}$ *which has some ergodic equilibrium state* ρ' *such that* $\lim_{|i| \to \infty} \rho'(A\tau_i B)$ *does not exist. Thus translation invariance is broken for* $\tilde{\Phi}$.

Proof. Taking ε sufficiently small in Corollary V.3.1, we obtain an invariant equilibrium state $\tilde{\rho}$ for some $\tilde{\Phi} \in \Phi_0 + \mathfrak{F}$ such that $\tilde{\rho}(A\tau_i B)$ does not tend to a limit. By the Lebesgue Dominated Convergence Theorem,

$\lim_{|i|\to\infty} \rho'(A\tau_i B)$ must fail to exist for some ergodic state ρ' in the ergodic decomposition of $\tilde{\rho}$. But if ρ' had short range correlations, $\rho'(A\tau_i B)$ would have the limit $\rho'(A)\rho'(B)$ as $|i| \to \infty$. ■

In many cases we can obtain more detailed results than those given in Theorems V.3.2 and V.3.3. For example, in the presence of a suitable symmetry we will be able to take $h = 0$ in (2). Moreover, in many cases it is useful to calculate the estimate on $|||\tilde{\Phi} - \Phi_0|||$ that is obtained. This is most conveniently done when the observable A of Theorem V.3.2 depends on only a single site. In some cases, our estimate agrees with that given by the mean field theory.

COROLLARY V.3.4. *Let* $A \in C(\Omega_0)$ *be real-valued and not identically zero, and suppose its distribution is symmetric with respect to the a priori measure* μ_0 *(so that* $\langle A^n \rangle_0 = 0$ *for* n *odd). Let* $\mathfrak{F}$ *be the cone of interactions* $\Psi \in \mathfrak{B}$ *of the form*

$$\text{(3)}\qquad \begin{aligned} &\Psi\{i,j\} = -J(i-j)(\tau_i A)(\tau_j A) \quad \textit{for } i \neq j,\ J \geq 0 \\ &\Psi(X) = 0 \quad \textit{otherwise.} \end{aligned}$$

Then for any $\delta > 0$ *there is* $\tilde{\Phi} \in \mathfrak{F}$ *with* $\sum_{i \neq 0} J(i) < \langle A^2 \rangle_0^{-1} + \delta$ *which exhibits "spontaneous magnetization," i.e., some invariant equilibrium state* ρ' *has* $\rho'(A) \neq 0$.

Note: The existence of a spontaneous magnetization implies that there is more than one ergodic state, since it is easily seen that any limit of Gibbs states for periodic boundary conditions will give A the expectation 0. The obvious quantum version of this result is still "essentially" classical because all the terms in the interaction commute. At some cost in simplicity we could obtain a similar result for "genuinely" quantum interactions of the form

$$\Psi\{i,j\} = -J(i-j)(\tau_i \vec{A}) \cdot (\tau_j \vec{A})$$

where $\vec{A} = (A_1, \cdots, A_k)$ is a k-tuple of self-adjoint operators. We will consider one case of this form, the Heisenberg model, below.

Proof. For $h > 0$ let ρ_h be the Gibbs state for an "external field" interaction $\Psi_h\{i\} = -h\tau_i A$. Thus ρ_h is the product measure which at each site is $\langle e^{hA}\rangle_0^{-1} e^{hA}\mu_0$. We will take $\Phi_0 = 0$; note $P(0) = 0$. Applying Corollary V.3.1 with $\varepsilon < \rho_h(A)^2 \|A\|^{-2}$, we obtain $\tilde{\Phi} \in \mathfrak{F}$ with

$$\sum_{i \neq 0} J(i) = \frac{2|||\tilde{\Phi}|||}{\|A\|^2} \leq \frac{-2\, s(\rho_h)}{\varepsilon \|A\|^2}$$

and an invariant equilibrium state $\tilde{\rho}$ with $\tilde{\rho}(A\tau_i A) \geq \rho_h(A)^2 - \varepsilon \|A\|^2 > 0$. Thus some ρ' in the ergodic decomposition of $\tilde{\rho}$ must have $\rho'(A) \neq 0$. If R is the infimum of $\sum_{i \neq 0} J(i)$ over all interactions in $\mathfrak{F}$ which exhibit spontaneous magnetization, this shows $R \leq -2\, s(\rho_h)\rho_h(A)^{-2}$. As $h \to 0$ we have

$$\rho_h(A) = \frac{\langle Ae^{hA}\rangle_0}{\langle e^{hA}\rangle_0} = h\langle A^2\rangle_0 + O(h^2)$$

$$-s(\rho_h) = \left\langle \frac{e^{hA}}{\langle e^{hA}\rangle_0} \ln \frac{e^{hA}}{\langle e^{hA}\rangle_0} \right\rangle_0 = \frac{h\langle Ae^{hA}\rangle_0}{\langle e^{hA}\rangle_0} - \ln \langle e^{hA}\rangle_0$$

$$= \frac{h^2}{2}\langle A^2\rangle_0 + O(h^3)$$

so that $R \leq \langle A^2\rangle_0^{-1}$. ∎

The mean field theory for this situation is as follows: the site j is acted on by a "magnetic field" $\sum_{i \neq j} J(i-j)\tau_i A$. We replace the variable $\tau_i A$ by its mean $\rho(A)$ for an invariant state ρ. Now we have a "mean field" acting on each site, so the equilibrium state should be ρ_h as above with $h = J\rho(A)$, where $J = \sum_{i \neq 0} J(i)$. For consistency we should have $\rho = \rho_h$, i.e., $h = J\,\rho_h(A)$. The mean field theory predicts a spontaneous magnetization when this equation has a solution $h \neq 0$. For large h we have $J\,\rho_h(A) \leq J\|A\| < h$, and so there will be a nonzero solution if $J \frac{d}{dh} \rho_h(A)|_{h=0} > 1$. Since $\rho_h(A) = h\langle A^2\rangle_0 + O(h^2)$, the "critical" J is at most $\langle A^2\rangle_0^{-1}$. which agrees with Corollary V.3.4.

In some cases, including the Ising model, $\rho_h(A)$ will be a concave function of $h > 0$, so that $\langle A^2\rangle_0^{-1}$ will actually be the critical J for the mean field theory. In other cases the mean field theory may have a lower critical J. We may also obtain a better estimate on R by the method of Corollary V.3.4 by not taking $h \to 0$ in some cases. In general the estimate on R we obtain will be different from the critical J of the mean field theory for these cases. In the case of the Ising model, where our estimate $R \leq 1$ agrees with the mean field value, we can in fact say $R = 1$: Griffiths [11] showed, using correlation inequalities, that an Ising ferromagnet with $\sum_{i \neq 0} \tanh J(i) < 1$ (and thus *a fortiori* one with $\sum_{i \neq 0} J(i) < 1$) has no spontaneous magnetization.

We now consider the (quantum) isotropic Heisenberg model. This is an important example, about which much less is known than about the Ising model. The Ising model has a spontaneous magnetization in one dimension for ferromagnetic interactions decaying as $|i-j|^{-r}$ with $1 < r < 2$ [5], and in two or more dimensions for nearest-neighbor interactions. In contrast, for the isotropic Heisenberg model there are no translation-invariant interactions in any dimension for which a spontaneous magnetization has been proven to exist. While our methods are unable to actually exhibit such an interaction, they do provide an existence proof.

We take the Hilbert space $\mathcal{H}_i$ at each site $i \in Z^\nu$ to be two-dimensional, with the usual Pauli matrices σ_i^x, σ_i^y, σ_i^z. An *isotropic Heisenberg interaction* is one of the form

$$\Psi_J\{i,j\} = -J(i-j)\,\vec{\sigma}_i \cdot \vec{\sigma}_j \quad \text{for } i \neq j$$

$$\Psi_J(X) = 0 \quad \text{if } |X| \neq 2 .$$

It is *ferromagnetic* if all $J(i) \geq 0$, *antiferromagnetic* if all $J(i) \leq 0$. An *external field* is a one-body interaction of the form

$$\Psi_{\vec{h}}\{i\} = \vec{h} \cdot \vec{\sigma}_i \quad \text{with} \quad \vec{h} \in R^3 .$$

Only a slight modification of our methods is needed to deal with $\vec{\sigma}_i \cdot \vec{\sigma}_j$ rather than an ordinary product of operators. First we will obtain an analogue of Theorem V.3.2 (without the use of symmetry). Here we let $\mathfrak{F}$ be the cone of interactions $\Psi_J + \Psi_{\vec{h}}$ where $\vec{h} \in R^3$ and $\Psi_J \in \mathcal{B}$ is a ferromagnetic isotropic Heisenberg interaction. We can take $\rho_0 = \frac{1}{2}(\rho_+ + \rho_-)$, where $\rho_{\pm}$ are the states in which all spins in the z direction are "up" and "down" respectively $(\rho_{\pm}(\sigma_i^z) = \pm 1)$. By Corollary V.3.1, for any $\Phi_0 \in \mathcal{B}$ and $\varepsilon > 0$ there is an invariant equilibrium state $\tilde{\rho}$ for some interaction $\tilde{\Phi} = \Phi_0 + \Phi_J + \Phi_h \in \Phi_0 + \mathfrak{F}$ with

$$|\tilde{\rho}(\vec{u} \cdot \vec{\sigma}_i)| \leq \varepsilon |\vec{u}| \quad \text{for all } \vec{u} \in R^3 \quad (|\vec{u}| \text{ is the Euclidean norm})$$

$$\tilde{\rho}(\vec{\sigma}_i \cdot \vec{\sigma}_j) \geq 1 - 3\varepsilon \quad \text{for all } i \neq j \in Z^\nu$$

and

$$|||\tilde{\Phi} - \Phi_0||| = |h| + \frac{3}{2} \sum_{i \neq 0} J(i) \leq \varepsilon^{-1}(P(\Phi_0) + \rho_0(A_{\Phi_0}) - s(\rho_0))$$

(note that $\|\vec{\sigma}_i \cdot \vec{\sigma}_j\| = 3$). We have

$$\tilde{\rho}(c_n(\sigma_i^x)^2 + c_n(\sigma_i^y)^2 + c_n(\sigma_i^z)^2) \geq 1 - 3\varepsilon .$$

Thus if $\varepsilon \leq \frac{1}{4}$ we obtain $\lim_{n \to \infty} \tilde{\rho}(c_n(\sigma_i^a)^2) > \tilde{\rho}(\sigma_i^a)^2$ where a is one of x, y and z, and some ergodic states in the decomposition of $\tilde{\rho}$ differ on σ_i^a.

If Φ_0 contains only terms which are even polynomials in the spins, we can dispense with the external fields: the conditions $\tilde{\rho}(\vec{\sigma}_i \cdot \vec{\sigma}_j) \geq 1 - 3\varepsilon > 0$ alone will ensure that there is a phase transition. Here the appropriate symmetry is known as *time reversal*. It is a *-antiautomorphism T of $\mathfrak{A}$ (i.e., an invertible positivity-preserving linear map with $T(AB) = T(B)T(A)$), and is determined by $T(\vec{\sigma}_i) = -\vec{\sigma}_i$. If $\Phi \in \mathcal{B}$ is invariant under T (which means each $\Phi(X)$ is an even polynomial in the spins σ_i^a, $i \in X$, $a = x,y,z$) and ρ is an invariant equilibrium state for Φ, then so is $\rho \circ T$. If $\rho(\vec{u} \cdot \vec{\sigma}_i) > 0$ for some $\vec{u} \in R^3$, then $\rho \cdot T(\vec{u} \cdot \vec{\sigma}_i) < 0$. Thus this "spontaneous magnetization" implies that there is more than one invariant equilibrium state.

COROLLARY V.3.5. *If* $\Phi_0 \in \mathcal{B}$ *is invariant under* T, *there is a ferromagnetic isotropic Heisenberg interaction* Φ_J *such that* $\Phi_0 + \Phi_J$ *exhibits spontaneous magnetization* ($\tilde{\rho}(\vec{u} \cdot \vec{\sigma}_i) \neq 0$ *for some* $\vec{u} \in R^3$ *and some invariant equilibrium state* $\tilde{\rho}$). *If* $\Phi_0 = 0$, *then for any* $\delta > 0$ *there is such a* Φ_J *with* $\sum_{i \neq 0} J(i) \leq 1 + \delta$.

Proof. For use in the second part, we take $\rho_{\vec{h}}$ as the equilibrium state for an external field $-\vec{h}$ (for the first part, we could use the same ρ_0 as previously). Taking this time $\mathfrak{F}$ as the cone of ferromagnetic isotropic Heisenberg interactions in $\mathcal{B}$, we will obtain for ε sufficiently small $\tilde{\rho}(\vec{\sigma}_i \cdot \vec{\sigma}_j) \geq c > 0$ for some c and all $i, j \in Z^\nu$. Some ergodic state in the decomposition of $\tilde{\rho}$ will then satisfy

$$\rho'(\sigma_i^x)^2 + \rho'(\sigma_i^y)^2 + \rho'(\sigma_i^z)^2 = \lim_{n \to \infty} \rho'(c_n(\vec{\sigma}_i) \cdot c_n(\vec{\sigma}_i)) \geq c > 0 .$$

Thus we have $\rho'(\vec{u} \cdot \vec{\sigma}_i) > 0$ where $\vec{u} = (\rho'(\sigma_i^x), \rho'(\sigma_i^y), \rho'(\sigma_i^z))$. For a given $\vec{h}$ (say in the z direction for convenience), we can take any $\varepsilon < \frac{1}{3}\rho_{\vec{h}}(\vec{\sigma}_i) \cdot \rho_{\vec{h}}(\vec{\sigma}_j) = \frac{1}{3}\rho_{\vec{h}}(\sigma_i^z)^2$; just as in Corollary V.3.4 we have $R \leq -2\, s(\rho_{\vec{h}})\rho_{\vec{h}}(\sigma_i^z)^{-2}$, where R is the infimum of $\sum_{i \neq 0} J(i)$ for ferromagnetic isotropic Heisenberg interactions with spontaneous magnetization. Since we are now dealing only with external fields and spins in the z direction, our result will be the same as for the Ising model: $\rho_{\vec{h}}(\vec{\sigma}_i) = \vec{h} + O(|h|^2)$ and $s(\rho_{\vec{h}}) = -\frac{|h|^2}{2} + O(|h|^3)$ so that $R \leq 1$. ∎

The "conventional wisdom" is that a phase transition is more difficult to obtain for a Heisenberg model than for the corresponding Ising model. Thus we would expect there to be no spontaneous magnetization in the Heisenberg ferromagnet when $\sum_{i \neq 0} J(i) < 1$, just as for the Ising ferromagnet. However, as yet this is only a conjecture. The best lower bound for R that is known rigorously is due to Greenberg [10]: there is no spontaneous magnetization if $\sum_{i \neq 0} J(i) < 0.025$.

Another interesting point is that we can obtain a phase transition with an interaction in which most of the J(i) are zero. In fact, for any infinite subset S of the lattice, we can restrict our cone $\mathfrak{F}$ to those (ferromagnetic isotropic Heisenberg) interactions where $J(i) = 0$ when $i \notin S$. We then obtain an interaction Φ_J in this cone with an invariant equilibrium state $\tilde{\rho}$ such that $\tilde{\rho}(\vec{\sigma}_0 \cdot \vec{\sigma}_i) \geq c > 0$ for all $i \in S$. Now it is not clear whether this implies the existence of more than one invariant equilibrium state, or a spontaneous magnetization, but it does mean that $\frac{1}{2}(\tilde{\rho} + \tilde{\rho} \cdot T)$ lacks short range correlations, so there is more than one KMS state. On the other hand, for a similar situation in the Ising model we can apply the results of Lebowitz and Martin-Löf [35]: if an Ising ferromagnet has more than one DLR state, then it has spontaneous magnetization.

Finally, we will examine a case of "antiferromagnetic" breaking of translation invariance in the Heisenberg model. Note that in the Ising model we can change a ferromagnet to an antiferromagnet by reversing the spins in a sublattice; however, this is impossible for the Heisenberg model, where there can be no symmetry taking $\vec{\sigma}_i \cdot \vec{\sigma}_j$ to $-\vec{\sigma}_i \cdot \vec{\sigma}_j$ (the two have different spectra). We will consider the lattice Z^ν to be divided into an A and a B sublattice, such that A is a subgroup of index 2 and B is the other coset (e.g., A and B may consist of sites i with $i_1 + \cdots + i_\nu$ even and odd respectively).

COROLLARY V.3.6. *For any* $\Phi_0 \in \mathcal{B}$ *there is an isotropic Heisenberg interaction* $\Phi_J \in \mathcal{B}$ *with* $J(i) = 0$ *for* $i \in A$ *and* $J(i) \leq 0$ *for* $i \in B$, *such that* $\Phi_0 + \Phi_J$ *has an invariant equilibrium state* $\tilde{\rho}$ *with* $\sup_{i \in B} \tilde{\rho}(\vec{\sigma}_0 \cdot \vec{\sigma}_i) < 0$, *and thus exhibits broken translation invariance. If* $\vec{h} \in R^3$ *let* $R(\vec{h})$ *be the infimum of* $-\sum_{i \in B} J(i)$ *over interactions* Φ_J *of the above form for which* $\Phi_{\vec{h}} + \Phi_J$ *exhibits this broken translation invariance. Then* $R(\vec{h}) \leq 2 \ln(2 \cosh |\vec{h}|)$ *and* $R(0) \leq 1$.

Proof. Let ρ_a be the state with $\rho_a(\sigma_i{}^z) = +1$ for $i \in A$ and -1 for $i \in B$, and let ρ_0 be the translation-invariant state $\frac{1}{2}(\rho_a + \rho_a \circ T)$ (which is $\frac{1}{2}(\rho_a + \rho_a \circ \tau_j)$ for $j \in B$). We have $\rho_0(\vec{u} \cdot \vec{\sigma}_0) = 0$ for all $\vec{u} \in R^3$, $\rho_0(\vec{\sigma}_0 \cdot \vec{\sigma}_j) = -1$ for $j \in B$, and $s(\rho_0) = -\ln 2$. The methods of Corollary V.3.1 apply as usual; the condition $\sup_{i \in B} \tilde{\rho}(\vec{\sigma}_0 \cdot \vec{\sigma}_i) < 0$ implies the existence of broken translation invariance, because since

$$\lim_{n \to \infty} \tilde{\rho}(c_n(\sigma_0{}^x)^2 + c_n(\sigma_0{}^y)^2 + c_n(\sigma_0{}^z)^2) \geq 0$$

we must have $\limsup_{|i| \to \infty} \tilde{\rho}(\vec{\sigma}_0 \cdot \vec{\sigma}_i) \geq 0$. In the case where $\Phi_0 = \Phi_{\vec{h}}$ is an external field, we have $P(\Phi_{\vec{h}}) = \ln \cosh |\vec{h}|$ and $\rho_0(A_{\Phi_{\vec{h}}}) = 0$, so that $R(\vec{h}) \leq 2 \ln(2 \cosh |\vec{h}|)$. For $\vec{h} = 0$ we can do better by replacing ρ_a with the equilibrium state for an alternating external field that is $+\vec{h}'$ for A sites and $-\vec{h}'$ for B sites, and taking $\vec{h}' \to 0$. This yields $R(0) \leq 1$, just as for the ferromagnet. ∎

VI. THE GIBBS PHASE RULE

VI.1. *Baire category, Hausdorff dimension and the phase rule*

By a theorem of Mazur ([19], Satz 2), a continuous convex function on a real separable Banach space has unique tangent functionals on a dense G_δ set of points. This was applied to statistical mechanics by Gallavotti and Miracle-Solé [8] and Ruelle [24], yielding a "weak Gibbs Phase Rule": in any Banach space of interactions dense in $\mathcal{B}$ and containing $\mathcal{B}_0$ (the finite-range interactions), a dense G_δ set of interactions have unique invariant equilibrium states. In this chapter we prove a stronger version of the Gibbs Phase Rule, with, in a sense, much of the content of the original "thermodynamic" version. The qualification "in a sense" refers to our lack of analyticity properties that are assumed in thermodynamics; thus where the original version of the Phase Rule says "lines," our version will have sets which may look quite wild, or even be everywhere dense, but will be one-dimensional in a certain sense: namely, Hausdorff dimension.

Consider a chemical system where the temperature and the chemical potentials of N chemical components can be varied independently at equilibrium. This determines an N+1-dimensional space of interactions. The usual statement of the Gibbs Phase Rule in this situation is that when there are M pure phases there are N+2–M degrees of freedom remaining; that is, M pure phases coexist on an (N+2–M)-dimensional subset (presumably a manifold with boundary) of the N+1-dimensional space. Thus for a single-component system we have a two-dimensional phase diagram with perhaps some triple points, lines of coexistence of two phases, and unique phases over all the rest of the diagram.

For mathematical rigor, we must modify this statement somewhat: it should only be expected to hold "generically." Thus for a one-component system (with only finitely many chemicals available in the real world) we would not expect four phases to meet at one point on the phase diagram. But as mathematicians, we are allowed to vary the parameters of the interaction, and it is presumably possible in some cases to make two triple points collide, forming a quadruple point. In fact, symmetry considerations can cause the rule to be violated (although it may be a debatable point whether two phases related by a symmetry should really be considered as different). For example, consider a chemical substance having right-handed and left-handed stereoisomers. Interconversion between the two forms (a process called *racemization*) can occur (although for some substances this happens very slowly), so there is just one chemical component involved. In solution or in an amorphous solid, symmetry requires that the two forms have equal concentration at equilibrium. Crystals, on the other hand, tend to be virtually 100% right-handed or left-handed, because molecules of the same "handedness" fit together much better in a crystal lattice than those of different "handedness." Thus two crystalline phases, one right-handed and one left-handed, will coexist where the Phase Rule predicts only one phase. If, on the other hand, the symmetry could be removed (say by adding a right-handed chemical which is not allowed to racemize), we would expect only one phase to remain.

The other necessary modification, as we have mentioned, concerns the notion of dimension, and is necessitated by a lack of analyticity properties. Although for soluble models and probably in nature the thermodynamic functions and loci of phase transitions are all piecewise analytic, we are a long way from proving such results in general (there are some cases – high temperature, low activity, a ferromagnet with external field, and one-dimensional systems with short-range interactions – for which phase transitions do not occur and analyticity can be proven; see [25]). In any case, because of the pathologies we found in Section V.2, such results are only to be expected in $\tilde{\mathcal{B}}$ or $\mathcal{B}_r$ or even smaller spaces of interactions. Therefore we must use a definition of "dimension" of a set which applies to sets which may bear little resemblance to manifolds. The most suitable definition turns out to be that of Hausdorff dimension. For the reader's convenience, we have placed the definitions and essential facts of Hausdorff measure and dimension in Appendix A.

Another consequence of the lack of analyticity properties is that, although we can define a "phase" for a given interaction, we can not think of a phase *extending over a region of interactions*, with properties that continue analytically from one interaction to another. Thus in the thermodynamic formulation, a line of two coexisting phases in a two-dimensional diagram will have a definite phase on each side (e.g., liquid on one side, gas on the other); in our formulation, even if our one-dimensional set is a recognizable line, no connection will be made between the interactions on one side of it; it will simply be a set of points, at each of which there are two ergodic equilibrium states.

Like the theorem of Mazur, our result will be a general theorem about convex functions on Banach spaces, and we will state it in that context. Apart from the convexity and continuity of the pressure, all we will use from statistical mechanics is the *identification of the "number of pure phases"* for an interaction as one plus the dimension of the set of tangent functionals to the pressure at that interaction.

Consider a real separable Banach space $\mathfrak{X}$ with a continuous convex function P defined on a convex open subset D_P of $\mathfrak{X}$. For each $\Phi \in D_P$ we let Δ_Φ denote the set of tangent functionals to P at Φ. Let

$$F^m = \{\Phi \in D_P : \dim \Delta_\Phi \geq m\} \text{ for each } m \geq 0 .$$

Of course $F^0 = D_P$ since the Hahn-Banach Theorem shows there is at least one tangent functional at Φ. In our application to statistical mechanics, $D_P = \mathfrak{X}$ is one of our spaces of interactions, P is the pressure, and F^m is the set of interactions with more than m pure phases.

A k-dimensional subset $\mathfrak{M}$ (which for our purposes will be a linear subspace with perhaps a lower-dimensional subspace removed) of $\mathfrak{X}$ will be said to *satisfy the Phase Rule* if $F^m \cap M$ is empty for all $m > k$, and has Hausdorff dimension at most $k-m$ for $m \leq k$.

In order to speak of "generic" subspaces of $\mathfrak{X}$ in the Baire category sense, we need to make the set $\mathcal{G}^k$ of k-dimensional linear subspaces of $\mathfrak{X}$ into a complete metric space. This is done by using the metric

$$(1)\quad \begin{aligned} d(\mathfrak{M},\mathfrak{N}) = \inf\{r>0: \mathfrak{M}\cap B_1 \subseteq (\mathfrak{N}\cap B_1) + B_r \text{ and} \\ \mathfrak{N}\cap B_1 \subseteq (\mathfrak{M}\cap B_1) + B_r\} \end{aligned}$$

where B_s denotes the open ball of radius s about 0. It is not difficult to show that $\mathcal{G}^k$ is complete in this metric. Moreover, if a sequence of k-tuples $\vec{\Phi}_n = (\Phi_n^1, \cdots, \Phi_n^k)$ of points of $\mathfrak{X}$ tends to a k-tuple $(\Phi^1, \cdots, \Phi^k)$ which is linearly independent, then the subspaces spanned by $\{\Phi_n^1, \cdots, \Phi_n^k\}$ tend in this metric to the subspace spanned by $\{\Phi^1, \cdots, \Phi^k\}$. Thus the notions of "generic k-dimensional subspace" and "span of a generic k-tuple of points" can easily be seen to coincide. We will take some fixed linear subspace $\mathfrak{N}_0$ of dimension less than k. The set $\mathcal{G}_0^k = \{\mathfrak{N} \in \mathcal{G}^k : \mathfrak{N}_0 \subset \mathfrak{N}\}$ is closed in $\mathcal{G}^k$. Our main theorem, which will be Theorem VI.3.3, will state:

The set Q^k *of subspaces* $\mathfrak{N} \in \mathcal{G}_0^k$ *such that* $\mathfrak{N} \backslash \mathfrak{N}_0$ *satisfies the Phase Rule is a dense* G_δ *in* $\mathcal{G}_0^k$.

We can take the subspace $\mathfrak{N}_0$ to be $\{0\}$ to obtain a result about subspaces satisfying the Phase Rule except perhaps at 0 (in general, the function P may have a large set of tangent functionals at 0, so that the whole linear subspace could not satisfy the Phase Rule. In the case of statistical mechanics, however, there is a unique tangent functional at 0). Provision for a larger $\mathfrak{N}_0$ is made so that we can consider the following situation: suppose we start with a subspace $\mathfrak{N}_0$ which does not satisfy the Phase Rule. Our theorem says that generic higher-dimensional subspaces $\mathfrak{N}$ containing $\mathfrak{N}_0$ will have $\mathfrak{N}\backslash\mathfrak{N}_0$ satisfying the Phase Rule. Unless $\mathfrak{N}_0$ contains points with arbitrarily many pure phases, we can choose the dimension of $\mathfrak{N}$ large enough so that all of $\mathfrak{N}$ will then satisfy the Phase Rule. Thus if $\mathfrak{N}_0$ is a line which has a large set of points of two pure phases, it is contained in a plane whose intersection with F^1 is one-dimensional, as it should be for the Phase Rule. Note that any symmetry which might be associated with the failure of $\mathfrak{N}_0$ to satisfy the Phase Rule will be generically broken in taking $\mathfrak{N}$.

Note also that we can combine our Phase Rule with Corollary II.1.3 to obtain similar results on systems with additional symmetry. Suppose we have the symmetry group H in addition to translations (as described before Corollary II.1.3), with $\mathcal{B}^H$ the space of interactions which are invariant under H, and redefine a "pure phase" as an extremal (Z^ν and H)-invariant equilibrium state. We then find that generic finite-dimensional subspaces of $\mathcal{B}^H$ satisfy the Phase Rule with this new definition of pure phases. (We have not proven that the set E^H of states invariant under both Z^ν and H forms a Choquet simplex, but this is not really harder than the result for Z^ν alone; see [25], Corollary 6.4.2.)

VI.2. *Some point-set topology*

We will begin by studying the topological type of certain sets associated with convex functions and the Hausdorff measures of the intersections of linear subspaces with other sets. The result will be that the set Q^k, which we want to prove generic, is indeed a G_δ. It will also give us a collection of G_δ sets whose intersection is Q^k. The task of Section VI.3 will then be to show that the sets in this collection are all dense. Our first elementary lemma is actually unnecessary in the application to statistical mechanics, where by Theorems I.2.3 (for the classical case) and I.3.4 (for the quantum case) $|P(\Phi)-P(\Psi)| \leq |||\Phi-\Psi|||$.

LEMMA VI.2.1. P *is locally Lipschitz.*

Proof. Fix $\Phi \in D_P$ and take $\delta > 0$ so that $\|\Phi - \Psi\| \leq \delta$ implies $|P(\Phi) - P(\Psi)| < \frac{1}{2}$. Suppose $\|\Phi - \Psi\| < \frac{\delta}{4}$ and $\|\Phi - \Psi'\| < \frac{\delta}{4}$. We can write

$$\Psi' = \left(1 - \frac{2\|\Psi' - \Psi\|}{\delta}\right)\Psi + \frac{2\|\Psi' - \Psi\|}{\delta}\left(\Psi + \frac{\delta(\Psi' - \Psi)}{2\|\Psi' - \Psi\|}\right) \quad (\Psi' \neq \Psi).$$

Now since

$$\left|P(\Phi) - P\left(\Psi + \frac{\delta(\Psi' - \Psi)}{2\|\Psi' - \Psi\|}\right)\right| < \frac{1}{2}$$

we have

$$P(\Psi') \leq \left(1 - \frac{2\|\Psi' - \Psi\|}{\delta}\right) P(\Psi) + \frac{2\|\Psi' - \Psi\|}{\delta} P\left(\Psi + \frac{\delta(\Psi' - \Psi)}{2\|\Psi' - \Psi\|}\right)$$

$$\leq P(\Psi) + \frac{2\|\Psi' - \Psi\|}{\delta}.$$

Similarly, interchanging Ψ and Ψ', we obtain

$$P(\Psi) \leq P(\Psi') + \frac{2\|\Psi' - \Psi\|}{\delta}. \qquad \blacksquare$$

Now we introduce one-sided directional derivatives. For $\Phi \in D_P$ and $\Psi \in \mathcal{X}$, the function $t \mapsto t^{-1}(P(\Phi + t\Psi) - P(\Phi))$ is defined on some interval $(0, \varepsilon)$, increasing (by convexity) and bounded below: in fact if we take δ as in Lemma VI.2.1 then

$$\frac{P(\Phi + t\Psi) - P(\Psi)}{t} \geq -\frac{2\|\Psi\|}{\delta}.$$

Thus the one-sided directional derivative

$$h(\Phi, \Psi) \equiv \lim_{t \to 0+} \frac{P(\Phi + t\Psi) - P(\Phi)}{t}$$

exists, and the following properties are immediate:

(i) $h(\Phi, s\Psi) = s\, h(\Phi, \Psi)$ for $s \geq 0$.

(ii) $-h(\Phi, -\Psi) \leq h(\Phi, \Psi)$.

(iii) For fixed $\Phi \in D_P, h(\Phi,\cdot)$ is a continuous convex function on $\mathfrak{X}$.

(iv) For fixed Ψ, $h(\cdot,\Psi)$ is upper semicontinuous on D_P.

Suppose $\alpha \in \mathfrak{X}^*$ is tangent to P at $\Phi \in D_P$. Then for $t > 0$ sufficiently small (so $\Phi \pm t\Psi \in D_P$)

$$\frac{P(\Phi + t\Psi) - P(\Phi)}{t} \geq \alpha(\Psi) \geq \frac{P(\Phi) - P(\Phi - t\Psi)}{t}$$

so

$$h(\Phi,\Psi) \geq \alpha(\Psi) \geq -h(\Phi,-\Psi) .$$

Conversely, suppose $\mathfrak{M}$ is a linear subspace of $\mathfrak{X}$ and β is a linear functional on $\mathfrak{M}$ with $h(\Phi,\Psi) \geq \beta(\Psi) \geq -h(\Phi,-\Psi)$ for $\Psi \in \mathfrak{M}$. Then by the Hahn-Banach Theorem we can extend β to $\alpha \in \mathfrak{X}^*$ with $h(\Phi,\Psi) \geq \alpha(\Psi) \geq -h(\Phi,-\Psi)$ for all $\Psi \in \mathfrak{X}$. Thus $P(\Phi+\Psi) - P(\Phi) \geq \alpha(\Psi)$ whenever $\Phi + \Psi \in D_P$, i.e., α is tangent to P at Φ.

Let $\mathfrak{M}_\Phi \equiv \{\Psi \in \mathfrak{X}\colon h(\Phi,\Psi) = -h(\Phi,-\Psi)\}$ for each $\Phi \in D_P$. By the above remarks, $\mathfrak{M}_\Phi$ is exactly the set on which all members of Δ_Φ coincide. Consequently $\mathfrak{M}_\Phi$ is a closed linear subspace of $\mathfrak{X}$, and moreover $\dim \Delta_\Phi = \dim \mathfrak{X}/\mathfrak{M}_\Phi$.

LEMMA VI.2.2. *F^m is an F_σ set in $\mathfrak{X}$.*

Proof. Note that $\dim \mathfrak{X}/\mathfrak{M}_\Phi \geq m$ if and only if there is some m-dimensional linear subspace of $\mathfrak{X}$ intersecting $\mathfrak{M}_\Phi$ only at 0. Thus $\Phi \in D_P$ is in F^m if and only if there are $\Psi_1, \cdots, \Psi_m \in \mathfrak{X}$ such that $\sum_{j=1}^m c_j \Psi_j \notin \mathfrak{M}_\Phi$ for all $\vec{c} \in R^m \backslash \{0\}$. This means

$$h\left(\Phi, \sum_{j=1}^m c_j \Psi_j\right) + h\left(\Phi, -\sum_{j=1}^m c_j \Psi_j\right) > 0 .$$

By the continuity and homogeneity of $h(\Phi,\cdot), \Psi_1, \cdots, \Psi_m$ can be chosen so that

$$h\left(\Phi, \sum_{j=1}^{m} c_j \Psi_j\right) + h\left(\Phi, -\sum_{j=1}^{m} c_j \Psi_j\right) > 1 \text{ whenever } \sum_{j=1}^{m} c_j^2 = 1 .$$

Now let $\{\Psi_i\}$ be a dense sequence in $\mathfrak{X}$, and for each m-tuple $\vec{i} = (i_1, \cdots, i_m)$ of positive integers let

$$C_{\vec{i}} = \left\{ \Phi \in D_P : \forall \vec{c} \in S^{m-1},\ h\left(\Phi, \sum_{j=1}^{m} c_j \Psi_{i_j}\right) + h\left(\Phi, -\sum_{j=1}^{m} c_j \Psi_{i_j}\right) \geq 1 \right\} .$$

By the continuity of $h(\Phi, \cdot)$ we have $F^m = \bigcup_{\vec{i}} C_{\vec{i}}$. By the upper semicontinuity of $h(\cdot, \Psi)$ for each Ψ, $C_{\vec{i}}$ is closed in D_P, while D_P is an F_σ since it is open. Thus F^m is an F_σ. ■

We will actually use both Lemma VI.2.2 itself and the particular representation $F^m = \bigcup_{\vec{i}} C_{\vec{i}}$ from its proof. This will be important in reducing the problem to an essentially finite-dimensional one: to decide whether $\Phi \in C_{\vec{i}}$ we need only consider the directional derivatives $h(\Phi, \Psi)$ for Ψ in a certain finite-dimensional space

Recall that $\mathcal{G}_0^k$ is the set of k-dimensional subspaces of $\mathfrak{X}$ which contain a fixed subspace $\mathfrak{N}_0$ of lower dimension. $\mathcal{H}^r$ denotes r-dimensional Hausdorff measure (see Appendix A).

LEMMA VI.2.3. *Let* $C \subset \mathfrak{X}$ *be an* F_σ *set, and let* $r, t \geq 0$. *Then the set* $W = \{\mathfrak{N} \in \mathcal{G}_0^k : \mathcal{H}^r(\mathfrak{N} \cap C \setminus \mathfrak{N}_0) \leq t\}$ *is a* G_δ *in* $\mathcal{G}_0^k$.

Proof. First suppose C is closed, bounded and disjoint from $\mathfrak{N}_0$. Let $U \subset \mathfrak{X}$ be open and $\mathfrak{N} \in \mathcal{G}_0^k$ with $C \cap \mathfrak{N} \subset U$. If $C \subset B_R$ then the compact set $\mathfrak{N} \cap \bar{B}_R$ is disjoint from the closed set $C \setminus U$, so that the distance between them is strictly positive. Now for any $\mathfrak{M} \in \mathcal{G}_0^k$ with $d(\mathfrak{M}, \mathfrak{N}) < R^{-1} \operatorname{dist}(C \setminus U, \mathfrak{N} \cap \bar{B}_R)$ we have

$$\mathfrak{M} \cap B_R \subset (\mathfrak{N} \cap B_R) + B_s \quad \text{for some} \quad s < \text{dist}(C \setminus U, \mathfrak{N} \cap \bar{B}_R)$$

so that $\mathfrak{M} \cap B_R$ is disjoint from $C \setminus U$. Thus $\mathfrak{M} \cap C \subset U$, and so the set of $\mathfrak{N} \in \mathcal{G}_0^k$ with $\mathfrak{N} \cap C \subset U$ is open in $\mathcal{G}_0^k$. Let W_{MN} be the set of $\mathfrak{N} \in \mathcal{G}_0^k$ such that $\mathfrak{N} \cap C$ has size-$\left(\frac{1}{M}\right)$ approximating measure (see Appendix A) $m_{(1/M)}^r(\mathfrak{N} \cap C) < t + \frac{1}{N}$; this means that $\mathfrak{N} \cap C$ can be covered by a sequence of open sets U_j of diameters $d_j < \frac{1}{M}$ with $\sum_j d_j^r < t + \frac{1}{N}$. For each sequence U_j, the set of $\mathfrak{N} \in \mathcal{G}_0^k$ with $\mathfrak{N} \cap C \subset \bigcup_j U_j$ is open, so W_{MN} is also open. Thus $W = \bigcap_{MN} W_{MN}$ (by the definition of r-dimensional Hausdorff measure) is a G_δ.

Now in the general case (where C is no longer closed, bounded and disjoint from $\mathfrak{N}_0$) we can write $C \setminus \mathfrak{N}_0$ as the union of an increasing family of closed and bounded sets C_n. Then $W = \bigcap_n W_n$ where $W_n = \{\mathfrak{N} \in \mathcal{G}_0^k \colon \mathcal{H}^r(\mathfrak{N} \cap C_n) \leq t\}$.   ∎

Note that having Hausdorff dimension at most n is equivalent to having $\mathcal{H}^r$ measure zero for any sequence of reals $r > n$ decreasing to n. Moreover, a set is empty if and only if its $\mathcal{H}^0$ measure is zero ($\mathcal{H}^0$ is just counting measure). Thus Lemmas VI.2.2 and VI.2.3 tell us the following:

COROLLARY VI.2.4.   *The set* $Q^k = \{\mathfrak{N} \in \mathcal{G}_0^k \colon \mathfrak{N} \setminus \mathfrak{N}_0$ *satisfies the Phase Rule*} *is a* G_δ. *In fact, it is the intersection of a sequence of* G_δ *sets*

$$(2) \qquad W^n = \{\mathfrak{N} \in \mathcal{G}_0^k \colon \mathcal{H}^{s_n}(\mathfrak{N} \cap C^n \setminus \mathfrak{N}_0) = 0\}$$

with $s_n > k - m_n$, $C^n = \{\Phi \in D_P \colon \mathcal{Z}_n \cap \mathfrak{M}_\Phi = \{0\}\}$ *and* $\mathcal{Z}_n$ *an* m_n*-dimensional linear subspace of* $\mathcal{X}$.

VI.3. *Proof of the phase rule*

We now have to show that the sets W^n of formula (2) are dense in $\mathcal{G}_0^k$. The Baire Category Theorem will then complete the proof of our main theorem by showing that Q^k is dense. Our fundamental tool in this section is the following consequence of an inequality of Federer [6]; it is presented in the Appendix as Corollary A.5.

Let X *and* Y *be metric spaces,* Y *boundedly compact (i.e., closed balls in* Y *are compact), and let* $f: X \to Y$ *be locally Lipschitz. If* $A \subset X$ *has* σ*-finite* $\mathcal{H}^{r+s}$ *measure, then* $\{y \in Y: \mathcal{H}^s(A \cap f^{-1}\{y\}) > 0\}$ *has* σ*-finite* $\mathcal{H}^r$ *measure, and*

$$\{y \in Y: A \cap f^{-1}\{y\} \text{ has non-}\sigma\text{-finite } \mathcal{H}^s \text{ measure}\}$$

has $\mathcal{H}^r$ *measure zero. If* $\mathcal{H}^{r+s}(A) = 0$, *then*

$$\mathcal{H}^r\{y \in Y: \mathcal{H}^s(A \cap f^{-1}\{y\}) > 0\} = 0 .$$

Our first use of this result will deal with the case of finite-dimensional $\mathfrak{X}$. The following theorem is due to Anderson and Klee [1]; their proof used "upper semicontinuous collections" of sets instead of Corollary A.5.

THEOREM VI.3.1. *If* $\mathfrak{X}$ *has dimension* $n < \infty$, *then* F^m *has* σ*-finite* $\mathcal{H}^{n-m}$ *measure for all* $m \leq n$.

Proof. We will construct a contraction map f_1 on $\mathfrak{X}$ which sets up a one-to-one affine correspondence between Δ_Φ and $f_1^{-1}\{\Phi\}$ for each $\Phi \in D_P$. Thus Φ will be in F^m if and only if $\mathcal{H}^m(f_1^{-1}\{\Phi\}) > 0$. Corollary A.5 will then be applied: since $\mathfrak{X}$ has σ-finite $\mathcal{H}^n$ measure (which is equivalent to Lebesgue measure by Corollary A.6), F^m has σ-finite $\mathcal{H}^{n-m}$ measure.

We can give $\mathfrak{X}$ an inner product, denoted $\Phi \cdot \Psi$, which makes it (isomorphically) into a real Hilbert space. Moreover, we can assume $P > 0$ (otherwise replace P by $P - \alpha + c$ where α is a P-bounded linear functional and $c > \sup\{\alpha(\Phi) - P(\Phi): \Phi \in D_P\}$). Let $G(P)$ denote the graph of P (as a subset of the Hilbert space $\mathfrak{X} \oplus R$), and $\overline{G(P)}$ its closure (which is $G(P)$ itself if $D_P = \mathfrak{X}$). For $\Psi \in \mathfrak{X}$ let $f(\Psi) = (f_1(\Psi), f_2(\Psi))$ be the closest point to $(\Psi,0)$ in $\overline{G(P)}$. This is well-defined since $\mathcal{X}$ is finite-dimensional and P is convex and positive. Now for $\Psi_1,\Psi_2 \in \mathcal{X}$ and $0 \leq t \leq 1$ we have

$$\|f(\Psi_1)-(\Psi_1,0)\| \leq \|tf(\Psi_2)+(1-t)f(\Psi_1)-(\Psi_1,0)\|$$

since $tf(\Psi_2)+(1-t)f(\Psi_1)$ is on or above the graph of P. Thus

$$(f(\Psi_2)-f(\Psi_1)) \cdot (f(\Psi_1)-(\Psi_1,0)) \geq 0 .$$

Similarly

$$(f(\Psi_2)-f(\Psi_1)) \cdot ((\Psi_2,0)-f(\Psi_2)) \geq 0 .$$

Adding these, we obtain

$$\|f(\Psi_2)-f(\Psi_1)\|^2 \leq (f(\Psi_2)-f(\Psi_1)) \cdot (\Psi_2-\Psi_2,0)$$

$$\|\Psi_2-\Psi_1\| \geq \|f(\Psi_2)-f(\Psi_1)\| \geq \|f_1(\Psi_2)-f_1(\Psi_1)\| .$$

Thus f_1 is a contraction.

Now for $\Phi \in D_P$, $f_1(\Psi)=\Phi$ if and only if the hyperplane

$$\{(\Psi',t)\colon (\Psi'-\Phi,t-P(\Phi))\cdot(\Phi-\Psi,P(\Phi))=0\}$$

is tangent to $G(P)$ at $(\Phi,P(\Phi))$, i.e., if and only if the linear functional

$$\Psi' \mapsto \Psi' \cdot \frac{(\Psi-\Phi)}{P(\Phi)}$$

is tangent to P at Φ. This sets up the required one-to-one affine correspondence between Δ_Φ and $f_1^{-1}\{\Phi\}$. ∎

A second application of Corollary A.5 gives us control over the intersection of lower-dimensional subspaces with F^m (in the finite-dimensional case). It will also be of interest for what it says about lower-dimensional subspaces of subspaces satisfying the Phase Rule.

LEMMA VI.3.2. *Let* $\mathfrak{X}$ *have finite dimension* n, *and* $k > j = \dim \mathfrak{N}_0$. *Let* $F \subset \mathfrak{X}$ *have* σ*-finite* $\mathcal{H}^r$ *measure. Then for (Lebesgue) almost every* $\mathfrak{N} \in \mathcal{G}_0^k$, $F \cap \mathfrak{N} \backslash \mathfrak{N}_0$ *has* σ*-finite* $\mathcal{H}^{r+k-n}$ *measure when* $r+k \geq n$, *and is empty when* $r+k < n$.

Proof. First we will treat the simplest case, where $j = k-1$. Here we can define a map $g: \mathfrak{X} \backslash \mathfrak{N}_0 \to \mathcal{G}_0^k$ by $g(\Phi) = \text{span}\,\{\mathfrak{N}_0, \Phi\}$. This is easily seen to be locally Lipschitz (in fact smooth). Note that for $\mathfrak{N} \in \mathcal{G}_0^k$, $g^{-1}\{\mathfrak{N}\}$ $\mathfrak{N} \backslash \mathfrak{N}_0$. By Corollary A.5, the set

$$\{\mathfrak{N} \in \mathcal{G}_0^k : F \cap \mathfrak{N} \backslash \mathfrak{N}_0 \text{ has non-}\sigma\text{-finite } \mathcal{H}^{r+k-n} \text{ measure}\}$$

has $\mathcal{H}^{n-k}$ measure zero if $r+k \geq n$, while if $r+k < n$, F has $\mathcal{H}^{n-k}$ measure zero and

$$\{\mathfrak{N} \in \mathcal{G}_0^k : F \cap \mathfrak{N} \backslash \mathfrak{N}_0 \neq \emptyset \text{ (i.e., has positive } \mathcal{H}^0 \text{ measure)}\}$$

has $\mathcal{H}^{n-k}$ measure zero. Since $\mathcal{G}_0^k$ is an n–k-dimensional manifold in this case, we are done.

In the general case we can take the map $g(\Phi, \mathcal{Y}) = \text{span}\,\{\Phi, \mathcal{Y}, \mathfrak{N}_0\}$ from $\{(\Phi, \mathcal{Y}) \in (\mathfrak{X} \backslash \mathfrak{N}_0) \times \mathcal{G}^{k-j-1} : \mathcal{Y} \cap \text{span}\,\{\Phi, \mathfrak{N}_0\} = \{0\}\}$ to $\mathcal{G}_0^k$. Again this is locally Lipschitz, and the inverse image of $\mathfrak{N} \in \mathcal{G}_0^k$ is

$$g^{-1}\{\mathfrak{N}\} = \{(\Phi, \mathcal{Y}) \in (\mathfrak{N} \backslash \mathfrak{N}_0) \times \mathcal{G}^{k-j-1}(\mathfrak{N}) : \mathcal{Y} \cap \text{span}\,\{\Phi, \mathfrak{N}_0\} = \{0\}\}$$

where $\mathcal{G}^s(\mathfrak{N})$ denotes the set of s-dimensional linear subspaces of $\mathfrak{N}$. If $F \cap \mathfrak{N} \backslash \mathfrak{N}_0$ has non-σ-finite $\mathcal{H}^d$ measure, then $(F \times \mathcal{G}^{k-j-1}) \cap g^{-1}\{\mathfrak{N}\}$ has non-σ-finite $\mathcal{H}^{d+t}$ measure, where t is the dimension of the manifold $\mathcal{G}^{k-j-1}(\mathfrak{N})$. But $F \times \mathcal{G}^{k-j-1}$ has σ-finite $\mathcal{H}^{r+u}$ measure, where u is the dimension of $\mathcal{G}^{k-j-1}$. By Corollary A.5,

$$\mathcal{H}^{r+u-d-t}\{\mathfrak{N} \in \mathcal{G}_0^k : F \cap \mathfrak{N} \backslash \mathfrak{N}_0 \text{ has non-}\sigma\text{-finite } \mathcal{H}^d \text{ measure}\} = 0\,.$$

We take $d = r+k-n$ and note that $\mathcal{G}^{k-j-1}(\mathfrak{N})$, $\mathcal{G}^{k-j-1}$ and $\mathcal{G}_0^k$ have dimensions $t = (k-j-1)(j+1)$, $u = (k-j-1)(n-k+j+1)$, and $(n-k)(n-j)$ respectively. Therefore we find that if $r+k \geq n$, $F \cap \mathfrak{N}\backslash\mathfrak{N}_0$ has σ-finite $\mathcal{H}^{r+k-n}$ measure for almost every $\mathfrak{N} \in \mathcal{G}_0^k$. If $r+k < n$, a similar argument shows that $F \cap \mathfrak{N}\backslash\mathfrak{N}_0 = \emptyset$ for almost every $\mathfrak{N}$. ∎

In the application to statistical mechanics, Lemma VI.3.2 says that if a finite-dimensional subspace $\mathfrak{X}$ of $\mathfrak{B}$ satisfies the Phase Rule, then so does almost every lower-dimensional subspace of $\mathfrak{X}$ of each dimension (note that here 0 has a unique tangent to the pressure). In some ways, a "Lebesgue almost everywhere" result is more satisfying than a "generic" result alone; there are well-known examples of "generic" subsets of the interval with Lebesgue measure zero. In $\mathcal{G}_0^k$ itself, there is no natural measure (in the infinite-dimensional case), so we must be satisfied with a "generic" Phase Rule. This result for finite-dimensional subspaces, however, increases our confidence that a "typical" subspace really will satisfy the Phase Rule.

We are now ready to complete the proof of our main result.

THEOREM VI.3.3. *The set* Q^k *of subspaces* $\mathfrak{N} \in \mathcal{G}_0^k$ *such that* $\mathfrak{N}\backslash\mathfrak{N}_0$ *satisfies the Phase Rule is a dense* G_δ *in* $\mathcal{G}_0^k$.

Proof. As remarked at the beginning of this section, it is enough to show that the sets W^n of formula (2) are dense in $\mathcal{G}_0^k$. Fix m_n, s_n and $\mathcal{Z}_n$ as in Corollary VI.2.4. For any $\mathfrak{N} \in \mathcal{G}_0^k$, take $\mathcal{Y} = \mathfrak{N} + \mathcal{Z}_n$. It is enough to show $W^n \cap \mathcal{G}_0^k(\mathcal{Y})$ is dense in $\mathcal{G}_0^k(\mathcal{Y})$. Taking $F = F^{m_n}$ and $r = (\dim \mathcal{Y}) - m_n$ in Lemma VI.3.2 (and using Theorem VI.3.1), we obtain a subset of full measure in $\mathcal{G}_0^k(\mathcal{Y})$ contained in W^n. ∎

APPENDIX A.

HAUSDORFF MEASURE AND DIMENSION

This appendix is a brief review of the theory of Hausdorff measure and Hausdorff dimension, which we use in Chapter VI. The material is adapted from [6] and [22].

Hausdorff measures $\mathcal{H}^r$ of all nonnegative dimensions r (not necessarily integers) are defined on any metric space; from them one can obtain a Hausdorff dimension for any subset. This seems to be the most natural way of defining dimension in the context of a metric space, for sets which may be very badly behaved. In the "good" cases, e.g., on a manifold, Hausdorff measure of the appropriate dimension is equivalent to Lebesgue measure, and the Hausdorff dimension of the manifold is its usual dimension.

Let X be a metric space, and $r \geq 0$. For each $\delta > 0$ and $A \subset X$ we let $m_\delta^r(A)$ be the infimum of $\sum_{S \in G} (\text{diam } S)^r$ over all countable families G of nonempty closed sets of diameter less than δ which cover A. Here for $r = 0$ we make the convention $0^0 = 1$. Let $\mathcal{H}^r(A) = \lim_{\delta \to 0} m_\delta^r(A)$. Clearly $m_\delta^r(A) > m_\eta^r(A)$ for $\delta < \eta$, so the limit (possibly $+\infty$) exists. $\mathcal{H}^r$ is called *r-dimensional Hausdorff measure*, and m_δ^r the *size* δ *approximating measure*. Note that we could take families of open sets instead of closed sets, and obtain the same m_δ^r and $\mathcal{H}^r$.

Let F be a countable family of subsets of X. Then

$$m_\delta^r(\cup F) \leq \sum_{S \in F} m_\delta^r(S)$$

and this inequality carries over to the limit as $\delta \to 0$. This means that m_δ^r and $\mathcal{H}^r$ are *outer measures*.

A set A is said to be *measurable* for an outer measure μ if for every set B, $\mu(B) = \mu(A \cap B) + \mu(B \setminus A)$. *Carathéodory's Criterion* ([6], p. 75) states that all Borel sets are measurable for μ if and only if $\mu(A \cup B) = \mu(A) + \mu(B)$ for all sets A,B with $\operatorname{dist}(A,B) > 0$.

If $\operatorname{dist}(A,B) \geq \delta$, any covering of $A \cup B$ by sets of diameter less than δ can be divided into a covering of A and a covering of B with no sets in common, and so $m_\delta^r(A \cup B) = m_\delta^r(A) + m_\delta^r(B)$. Thus if $\operatorname{dist}(A,B) > 0$ we have $\mathcal{H}^r(A \cup B) = \mathcal{H}^r(A) + \mathcal{H}^r(B)$, so that all Borel sets are $\mathcal{H}^r$-measurable. However, in general there are non-m_δ^r-measurable Borel sets.

It is easy to show that for any outer measure μ and any increasing sequence of μ-measurable sets (E_n), $\mu(\bigcup_n E_n) = \sup_n \mu(E_n)$. For non-measurable sets, this might not be true. Now any subset of X is contained in a Borel set of equal m_δ^r or $\mathcal{H}^r$ measure, which can be obtained as the intersection of $\cup G^n$ for a sequence (G^n) of countable coverings of our original set by open sets. This is a very useful regularity property for $\mathcal{H}^r$; e.g., it implies that $\mathcal{H}^r(\bigcup_n E_n) = \sup_n \mathcal{H}^r(E_n)$ for an arbitrary increasing sequence of sets. We will require some similar properties for m_δ^r; this is much more difficult, because Borel sets need not be m_δ^r-measurable.

LEMMA A.1. *Let* $A \subset X$ *and* $a \in X$. *Then* $m_\delta^r(A) = \lim_{R \to \infty} m_\delta^r(A \cap \bar{B}_R(a))$ *where* $\bar{B}_R(a)$ *is the closed ball of radius* R *about* a.

Proof. Let $s = \lim_{R \to \infty} m_\delta^r(A \cap \bar{B}_R(a))$. It is clear that $m_\delta^r(A) \geq s$, and so it suffices to show $m_\delta^r(A) \leq s$ when $s < \infty$. For positive integers j, let $C_j = \{x \in A\colon 2(j-1)\delta \leq d(a,x) \leq 2j\delta\}$. Since the distance between two of these "shells" which are not adjacent is at least δ, we have $\sum_{k \in K} m_\delta^r(C_k) = m_\delta^r(\bigcup_{k \in K} C_k) \leq s$ for any finite set K of positive integers containing no two adjacent integers. Therefore $\sum_{k=1}^{\infty} m_\delta^r(C_k) < \infty$, and $m_\delta^r(A) \leq \lim_{n \to \infty} (m_\delta^r(A \cap \bar{B}_n(a)) + \sum_{k=n+1}^{\infty} m_\delta^r(C_k)) = s$. ■

To facilitate the study of m_δ^r on arbitrary increasing sequences of sets, we introduce the *Hausdorff metric* on the space $\mathfrak{F}$ of nonempty closed subsets of X. For $A,B \in \mathfrak{F}$ we define

$$d(A,B) = \sup \{\operatorname{dist}(x,B), \operatorname{dist}(y,A) : x \in A,\ y \in B\} .$$

It is easily verified that $d(\cdot,\cdot)$ is a metric on $\mathfrak{F}$, and that in this metric the function $A \to \operatorname{diam} A$ is continuous on $\mathfrak{F}$.

LEMMA A.2 (Blaschke Selection Theorem). *If* X *is compact, then so is* $\mathfrak{F}$.

Proof. Let (C_i) be a sequence in $\mathfrak{F}$. Let (V_n) be a countable base of open sets for X. Choose inductively for each n infinite subsequences of (C_i), taking at step n either all the members of the previous subsequence that intersect V_n, or (if that would not be infinite) all those that are disjoint from V_n. Let (C_{i_n}) be the diagonal subsequence, and let D be the set of all limit points of sequences (x_n) with $x_n \in C_{i_n}$. Then it is not difficult to see that D is closed and nonempty, and $C_{i_n} \to D$ in the Hausdorff metric. ■

THEOREM A.3. *Suppose* X *is boundedly compact, i.e., every closed ball in* X *is compact. Then if* (E_n) *is an increasing family of subsets of* X *and* $0 < \delta < \eta$, *we have* $m_\eta^r(\bigcup_n E_n) \le \sup_n m_\delta^r(E_n)$.

Proof. Let $E = \bigcup_n E_n$. By Lemma A.1, it suffices to consider the case where $E \subset B_R(a)$ for some $a \in X$ and $R > 0$. We can also assume that $\sup_n m_\delta^r(E_n) = L < \infty$.

First we consider the case $r = 0$. Here L is an integer, and for each n sufficiently large there are nonempty closed subsets $F_1^n, \cdots, F_L^n$ of $\overline{B}_R(a)$ with diameters less than δ and $E_n \subset \bigcup_{i=1}^{L} F_i^n$. By the Blaschke Selection Theorem we obtain a subsequence (n_j) of positive integers such that $F_i^{n_j} \to F_i^*$ for each i, where $F_1^*, \cdots, F_L^*$ are nonempty closed sets. Then $\operatorname{diam} F_i^* \leq \delta$ and $E \subset \bigcup_{i=1}^{L} F_i^*$, so that $m_\eta^0(E) \leq L$.

Now let $r > 0$, and let $\varepsilon > 0$ be given. For each n we choose nonempty closed sets F_i^n in $\overline{B}_R(a)$ with $E_n \subset \bigcup_{i=1}^{\infty} F_i^n$, $\operatorname{diam} F_i^n < \delta$, and $\sum_{i=1}^{\infty} (\operatorname{diam} F_i^n)^r < L+\varepsilon$, and order them so that $\operatorname{diam} F_i^n \geq \operatorname{diam} F_{i+1}^n$. Using the Blaschke Selection Theorem and taking another diagonal subsequence, we obtain a sequence (n_j) of positive integers and a sequence (F_i^*) of nonempty closed sets such that $F_i^{n_j} \to F_i^*$ as $j \to \infty$ for each i. Letting $d_i = \operatorname{diam} F_i^*$ we have $\delta \geq d_i \geq d_{i+1}$ and $\sum_{i=1}^{\infty} d_i \equiv \iota \leq L+\varepsilon$. Now, in contrast to the case $r = 0$, we can not say $E \subset \bigcup_{i=1}^{\infty} F_i^*$. Instead we will show that points which do not lie sufficiently close to some F_i^* are contained in a set of small $\mathcal{H}^r$ measure.

For each i we choose δ_i with $d_i < \delta_i < \eta$ and $\delta_i^r < d_i^r + 2^{-i}\varepsilon$. Let $V_i = \{\alpha \in X\colon \operatorname{dist}(\alpha, F_i^*) \leq \frac{1}{2}(\delta_i - d_i)\}$. Then for each i, $\operatorname{diam} V_i \leq \delta_i$ and $F_i^{n_j} \subset V_i$ for j sufficiently large. Moreover,

$$\sum_{i=1}^{\infty} (\operatorname{diam} V_i)^r \leq \sum_{i=1}^{\infty} \delta_i^r < \iota + \varepsilon .$$

Let $V = \bigcup_{i=1}^{\infty} V_i$. We claim $\mathcal{H}^r(E_n \setminus V) \leq L - \iota + 3\varepsilon$ for each n.

Given n and $\beta > 0$, choose M so that $\sum_{i>M} d_i^r < \varepsilon$ and $d_M < \beta$. Choose $N \in (n_j)$ so that $N > n$, $\operatorname{diam} F_M^N < \beta$, and for $i \leq M$, $F_i^N \subset V_i$ and $(\operatorname{diam} F_i^N)^r \geq d_i^r - 2^{-i}\varepsilon$. Thus we have $E_n \setminus V \subset \bigcup_{i>M} F_i^N$. But for $i > M$, $\operatorname{diam} F_i^N \leq \operatorname{diam} F_M^N < \beta$, so that

$$m_\beta^r(E_n \setminus V) \leq \sum_{i>M} (\operatorname{diam} F_i^N)^r \leq L + \varepsilon - \sum_{i=1}^{M} (\operatorname{diam} F_i^N)^r$$

$$\leq L + \varepsilon - \sum_{i=1}^{M} (d_i^r - 2^{-i}\varepsilon) < L - \iota + 3\varepsilon .$$

Taking $\beta \to 0$ we obtain $\mathcal{H}^r(E_n \setminus V) \leq L - \iota + 3\varepsilon$ as claimed.

Now by the regularity property of $\mathcal{H}^r$ we also have $\mathcal{H}^r(E \setminus V) \leq L - \iota + 3\varepsilon$. Since V is covered by the V_i, we have $m^r_\eta(V) \leq \sum_{i=1}^{\infty} (\text{diam } V_i)^r \leq \iota + \varepsilon$, so $m^r_\eta(E) \leq \mathcal{H}^r(E \setminus V) + m^r_\eta(V) \leq L + 4\varepsilon$. Taking $\varepsilon \to 0$, we finally obtain $m^r_\eta(E) \leq L$. ■

Next we prove a fundamental inequality relating Hausdorff measures of different dimensions. The basic principle involved is depicted in Figure 4. Here we consider a projection of a square onto a line. Two lower bounds on the $\mathcal{H}^1$ measure (i.e., total length) of the figure "A" can be obtained by (i) counting the number of points whose inverse images intersect A in a set of length at least 1, or (ii) finding the total length of the set of points whose inverse images intersect A.

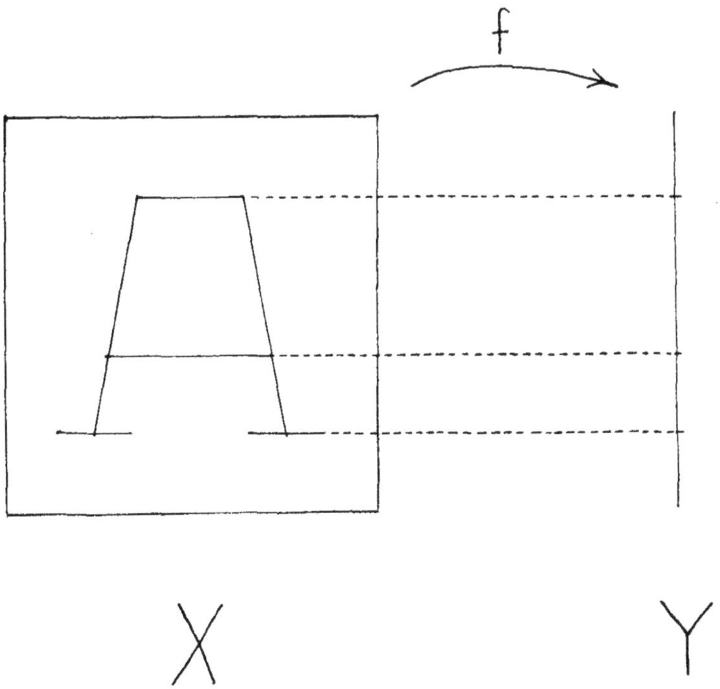

Figure 4

THEOREM A.4. *Let* X *and* Y *be metric spaces with* Y *boundedly compact,* $r,s \geq 0$, *and let* $f: X \to Y$ *be Lipschitz. Then there is a constant* C *such that for all* $A \subset X$, $t > 0$ *we have*

$$\mathcal{H}^r\{y \in Y: \mathcal{H}^s(A \cap f^{-1}\{y\}) > t\} \leq Ct^{-1}\mathcal{H}^{r+s}(A).$$

Proof. By a change of scale we can assume f has Lipschitz constant 1, so $\operatorname{diam} f(S) \leq \operatorname{diam} S$ for all $S \subset X$. We can also assume $\mathcal{H}^{r+s}(A) < \infty$.

For each j choose a countable cover G_j of A by nonempty closed sets of diameter less than j^{-1} such that $\sum_{S \in G_j} (\operatorname{diam} S)^{r+s} \leq m^{r+s}_{1/j}(A) + 1/j$.

Let $H_j = \{\overline{f(S)}: S \in G_j\}$, and for each $T \in H_j$ let $u_j(T) = \Sigma\{(\operatorname{diam} S)^s: S \in G_j, \overline{f(S)} = T\}$. Then since $A \cap f^{-1}\{y\} \subset \cup\{S \in G_j: y \in \overline{f(S)}\}$, we have

$$m^s_\eta(A \cap f^{-1}\{y\}) \leq \sum \{u_j(T): y \in T \in H_j\} \quad \text{for } \eta \geq j^{-1}.$$

Now we claim that

$$\mathcal{H}^r\{y \in Y: m^s_\eta(A \cap f^{-1}\{y\}) > t\} \leq 3^r t^{-1} \lim_{j \to \infty} \sum_{T \in H_j} u_j(T)(\operatorname{diam} T)^r.$$

Once the claim is proved, we can use the fact that $\{y.\ \mathcal{H}^s(A \cap f^{-1}\{y\}) > t\}$ is the increasing union of the sets $\{y: m^s_\eta(A \cap f^{-1}\{y\}) > t\}$ as $\eta \to 0$, and the estimate

$$u_j(T)(\operatorname{diam} T)^r \leq \sum \{(\operatorname{diam} S)^{r+s}: S \in G_j, \overline{f(S)} = T\}$$

to obtain

$$\mathcal{H}^r\{y: \mathcal{H}^s(A \cap f^{-1}\{y\}) > t\} \leq 3^r t^{-1} \lim_{j \to \infty} \sum_{S \in G_j} (\operatorname{diam} S)^{r+s} = 3^r t^{-1} \mathcal{H}^{r+s}(A).$$

To prove the claim, it suffices to show

$$m_\delta^r \left\{ y : \sum_{y \in T \in H_j} u_j(T) > t \right\} \leq 3^r t^{-1} \sum_{T \in H_j} u_j(T)(\operatorname{diam} T)^r \quad \text{for all } \delta > \frac{3}{j}.$$

By Theorem A.3 we need only verify this for finite subcollections H of H_j. Furthermore, we can approximate u_j by a function u on H taking rational values, approximate t by a rational number, and then multiply by the common denominator; this shows that we can assume that t and the values of u are positive integers. Now let $B = \{y: \sum_{y \in T \in H} u(T) > t\}$. We define inductively functions $v_0, \cdots, v_t$ on H and subfamilies $F_1, \cdots, F_t$ of H as follows: Let $v_0 = u$. For $1 \leq i \leq t$ choose F_i so that the members T of F_i are mutually disjoint with $v_{i-1}(T) \geq 1$ and $B \subset \cup\{T^*: T \in F_i\}$ where $T^* = \cup\{T' \in H: T' \cap T \neq \emptyset, \operatorname{diam} T' \leq \operatorname{diam} T\}$. Let

$$v_i(T) = \begin{cases} v_{i-1}(T) - 1 & \text{if } T \in F_i \\ v_{i-1}(T) & \text{otherwise}. \end{cases}$$

Then $B \subset \{y: \sum_{y \in T \in H} v_i(T) > t-i\}$, and so F_i can be chosen as above (take a set T_1 of maximal diameter from $\{T \in H: v_{i-1}(T) \geq 1\}$, then T_2 of maximal diameter from $\{T \in H: v_{i-1}(T) \geq 1, T \cap T_1 = \emptyset\}$, etc.).

Finally we find

$$t\, m_\delta^r(B) \leq \sum_{i=1}^{t} \sum_{T \in F_i} (\operatorname{diam} T^*)^r \leq \sum_{i=1}^{t} \sum_{T \in H} (v_{i-1}(T) - v_i(T))(3 \operatorname{diam} T)^r$$

$$\leq 3^r \sum_{T \in H} u(T)(\operatorname{diam} T)^r. \qquad \blacksquare$$

COROLLARY A.5. *Let* X *and* Y *be metric spaces,* Y *boundedly compact, and let* $f: X \to Y$ *be locally Lipschitz. If* $A \subset X$ *has* σ*-finite* $\mathcal{H}^{r+s}$ *measure, then* $\{y \in Y: \mathcal{H}^s(A \cap f^{-1}\{y\}) > 0\}$ *has* σ*-finite* $\mathcal{H}^r$ *measure, and*

$$\{y \in Y\colon A \cap f^{-1}\{y\} \text{ has non-}\sigma\text{-finite } \mathcal{H}^s \text{ measure}\}$$

has $\mathcal{H}^r$ *measure zero. If* $\mathcal{H}^{r+s}(A) = 0$, *then* $\{y \in Y\colon \mathcal{H}^s(A \cap f^{-1}\{y\}) > 0\}$ has $\mathcal{H}^r$ measure zero.

Proof. Write X as the increasing union of the sets

$$X_j = \{x \in X\colon \text{ for } z \text{ in some neighborhood of } x,\ d(f(x), f(z)) \le j\, d(x, z)\}\,.$$

Write A as the increasing union of sets $A_j \subset X_j$ with $\mathcal{H}^{r+s}(A_j) < \infty$. Then $\{y\colon \mathcal{H}^s(A \cap f^{-1}\{y\}) > 0\}$ is the increasing union of the sets $\{y\colon \mathcal{H}^s(A_j \cap f^{-1}\{y\}) > j^{-1}\}$, which have finite $\mathcal{H}^r$ measure, or zero $\mathcal{H}^r$ measure if $\mathcal{H}^{r+s}(A) = 0$. Moreover, $\{y\colon A \cap f^{-1}\{y\}$ has non-σ-finite $\mathcal{H}^s$ measure$\}$ is contained in the union of the sets $\{y\colon \mathcal{H}^s(A_j \cap f^{-1}\{y\}) = \infty\}$, which have $\mathcal{H}^r$ measure zero. ■

The above corollary is our most important tool in the proof of the Gibbs Phase Rule (Section VI.3). We will now use it to compare $\mathcal{H}^n$ to n-dimensional Lebesgue measure for positive integers n. (It is easy to see that $\mathcal{H}^0$ is counting measure.)

COROLLARY A.6. *On* $\mathbf{R}^n$, $\mathcal{H}^n$ *is a nonzero constant multiple of n-dimensional Lebesgue measure.*

Proof. Since $\mathcal{H}^n$ is clearly translation-invariant on $\mathbf{R}^n$, it suffices to show that $0 < \mathcal{H}^n(\mathbf{I}^n) < \infty$, where $\mathbf{I}^n$ is the unit n-cube in $\mathbf{R}^n$. To estimate the m_δ^r measure of $\mathbf{I}^n$, take $k > \sqrt{n}/\delta$ and cover $\mathbf{I}^n$ by k^n cubes of side k^{-1} and diameter $k^{-1}\sqrt{n}$, so that $m_\delta^n(\mathbf{I}^n) \le n^{n/2}$. Thus $\mathcal{H}^n(\mathbf{I}^n) < \infty$.

The proof that $\mathcal{H}^n(\mathbf{I}^n) > 0$ proceeds by induction. For $n = 1$ it is easy to show $\mathcal{H}^1(\mathbf{I}) = 1$. Suppose $\mathcal{H}^n(\mathbf{I}^n) > 0$. Let p be the first coordinate map from $\mathbf{R}^{n+1}$ to $\mathbf{R}$, so $\mathbf{I}^{n+1} \cap p^{-1}\{y\}$ is a unit n-cube for

$0 \leq y \leq 1$. Thus $\mathcal{H}^1\{y: \mathcal{H}^n(I^{n+1} \cap p^{-1}\{y\}) > 0\} = 1$, so that $\mathcal{H}^{n+1}(I^{n+1}) > 0$ by the last part of Corollary A.5. ■

In order to make $\mathcal{H}^n$ equal to Lebesgue measure on $\mathbf{R}^n$ it is common to multiply the $\mathcal{H}^r$ we have defined by a normalizing constant, which turns out to be $2^{-r}\pi^{r/2}/\Gamma(1+r/2)$.

We can use Hausdorff measures to define dimension in the following way: the *Hausdorff dimension* of the set A (in any metric space) is $\inf\{r \geq 0: \mathcal{H}^r(A) = 0\}$. Note that if $\mathcal{H}^r(A) < \infty$ then $\mathcal{H}^s(A) = 0$ for all $s > r$, and the same is true if A has σ-finite $\mathcal{H}^r$ measure. Thus Corollary A.6 implies that the Hausdorff dimension of $\mathbf{R}^n$ is n, as we would expect. Fractional Hausdorff dimensions also occur; for example, the standard Cantor set has Hausdorff dimension $\frac{\ln 2}{\ln 3}$.

Hausdorff dimension is a metric rather than topological concept, and is not invariant under arbitrary homeomorphisms. For example, Cantor sets of any desired Hausdorff dimension are easy to construct, and all are homeomorphic to each other. There is also a notion of *topological dimension* [15], defined inductively as follows: The empty set has dimension -1. X has dimension at most n if every point in X has arbitrarily small neighborhoods whose boundaries have dimension less than n. (Here n is always an integer.)

Topological dimension is really a matter of connectivity properties rather than "size" of a set. For example, the irrational numbers form a set of topological dimension 0. For our purposes, namely the Gibbs Phase Rule, this is not very satisfactory, and Hausdorff dimension is much more natural. Finally, we note that the Hausdorff dimension of a metric space is an upper bound on the topological dimension.

LEMMA A.7. *If* X *is a metric space and* $\mathcal{H}^{m+1}(X) = 0$, *then* X *has topological dimension at most* m. *Thus the Hausdorff dimension of* X *is at least the topological dimension.*

Proof. We use induction on m. If $\mathcal{H}^0(X) = 0$, then X is empty and has topological dimension -1. If $\mathcal{H}^{m+1}(X) = 0$ and $x \in X$, consider the function $d(x,\cdot)\colon X \to R$. By Corollary A.5, $\mathcal{H}^m\{y\colon d(x,y) = t\} = 0$ for almost every $t \in R^+$. By the induction hypothesis, for such t the topological dimension of $\{y\colon d(x,y) = t\}$ is at most $m-1$. Thus the topological dimension of X is at most m. ∎

APPENDIX B
CLASSICAL HARD-CORE CONTINUOUS SYSTEMS

Much of what has been done for lattice systems can be modified to deal with systems of classical particles with hard cores in Euclidean space. The hard core is useful because it prevents the presence of a very large number of particles in a small region. The analogous quantum-mechanical system is considerably more difficult to handle: for example, the variational principle has not been completely established [20].

Basic references for this appendix are [9], [18], and [21].

The simplest hard-core continuous system consists of identical point particles in $\mathbf{R}^2$ with a spherical hard core: the Euclidean distance between any two distinct particles must be at least the hard-core diameter 2. More generally, we may allow the particles some additional degrees of freedom, which may affect the hard-core radius. Our model might have several species of "molecules" with variable orientations and shapes.

We will suppose that the "internal" degrees of freedom of a particle are given by a compact metrizable set Q. A *configuration* ω then consists of a subset $\operatorname{spt}(\omega)$ of $\mathbf{R}^\nu$ (the "support" of ω), and a function $x \mapsto \omega(x)$ from $\operatorname{spt}(\omega)$ to Q. The restriction of a configuration ω to a subset Λ of $\mathbf{R}^\nu$ is the configuration $\omega_{\restriction\Lambda}$ with $\operatorname{spt}(\omega_{\restriction\Lambda}) = \operatorname{spt}(\omega) \cap \Lambda$ and $\omega_{\restriction\Lambda}(x) = \omega(x)$ for $x \in \operatorname{spt}(\omega_{\restriction\Lambda})$. We will say $\omega_1 \subseteq \omega_2$ if $\omega_1 = \omega_{2\restriction\operatorname{spt}(\omega_1)}$. The union $\bigcup_a \omega_a$ of a collection $\{\omega_a\}$ of configurations with disjoint supports is the configuration ω with $\operatorname{spt}(\omega) = \bigcup_a \operatorname{spt}(\omega_a)$ and $\omega(x) = \omega_a(x)$ if $x \in \operatorname{spt}(\omega_a)$. The translate of a configuration ω by $x \in \mathbf{R}^\nu$ will be denoted $\omega + x$. This is the configuration with $\operatorname{spt}(\omega+a) = \operatorname{spt}(\omega) + x$ and $(\omega+x)(y) = \omega(y-x)$.

A configuration ω is called *locally finite* if $\mathrm{spt}(\omega) \cap \Lambda$ is finite for every bounded set Λ. The space of locally finite configurations will be denoted $\Omega^{\ell f}$. This has a natural topology, which we will describe by presenting a basic system of neighborhoods for $\omega_0 \in \Omega^{\ell f}$. Let a metric d on Q be given. For $\varepsilon > 0$, the basic neighborhood $N_\varepsilon(\omega_0)$ consists of those $\omega \in \Omega^{\ell f}$ such that there are one-to-one maps ϕ and ϕ' from $\mathrm{spt}(\omega_0) \cap \{x: |x| < \frac{1}{\varepsilon}\}$ and $\mathrm{spt}(\omega) \cap \{x: |x| < \frac{1}{\varepsilon}\}$ into $\mathrm{spt}(\omega)$ and $\mathrm{spt}(\omega_0)$ respectively, with $|x - \phi(x)| < \varepsilon$ and $d(\omega\ (x), \omega(\phi(x))) < \varepsilon$ for $x \in \mathrm{spt}(\omega_0)$ with $|x| < \frac{1}{\varepsilon}$, and $|x - \phi'(x)| < \varepsilon$ and $d(\omega(x), \omega_0(\phi'(x))) < \varepsilon$ for $x \in \mathrm{spt}(\omega)$ with $|x| < \frac{1}{\varepsilon}$. This topology is metrizable: in fact we can take the metric $d(\omega, \omega') = \inf\{1, \varepsilon : \omega \in N_\varepsilon(\omega')\}$. This is not a complete metric, but we could complete it by allowing multiple occupation of points.

The hard core will be described by a continuous strictly positive function f on $Q \times Q \times S^{\nu-1}$ (where $S^{\nu-1}$ is the unit sphere in $\mathbf{R}^\nu$). The space Ω of allowed configurations consists of those $\omega \in \Omega^{\ell f}$ with $|x-y| \geq f(\omega(x), \omega(y), \frac{x-y}{|x-y|})$ for any two distinct $x, y \in \mathrm{spt}(\omega)$. It inherits the topology on $\Omega^{\ell f}$ introduced above. It is easy to prove that Ω is compact.

For any Borel subset Λ of $\mathbf{R}^\nu$, we will use $\Omega^{\ell f}_\Lambda$ and Ω_Λ to denote spaces of configurations in $\Omega^{\ell f}$ and Ω respectively with supports in Λ. If Λ is bounded we have $\Omega^{\ell f}_\Lambda = \bigcup_{n=0}^{\infty} G^{\ell f}_{n,\Lambda}$ where $G^{\ell f}_{n,\Lambda} = \{\omega \in \Omega^{\ell f}_\Lambda : |\mathrm{spt}(\omega)| = n\}$. We let $\mathcal{S}_\Lambda = \{A \times \Omega^{\ell f}_{\Lambda^c} : A \subset \Omega^{\ell f}_\Lambda \text{ Borel}\}$ be the σ-algebra of Borel sets in $\Omega^{\ell f}$ depending only on the configuration in Λ.

Let ν_0 be an "*a priori*" probability measure on Q. For any bounded Borel subset Λ of $\mathbf{R}^\nu$, a probability measure on $\Omega^{\ell f}_\Lambda$ is defined as follows: $\mu_0^\Lambda(\omega_\phi) = e^{-m(\Lambda)}$ (where $\omega_\emptyset$ is the empty configuration), while on $G^{\ell f}_{n,\Lambda}$ for $n \geq 1$, $\mu_0^\Lambda(d\omega) = \frac{e^{-m(\Lambda)}}{n!} dx_1 \cdots dx_n\, \nu_0(dq_1) \cdots \nu_n(dq_n)$ where $\mathrm{spt}(\omega) = \{x_1, \cdots, x_n\}$, $\omega(x_j) = q_j$, and $m(\cdot)$ is Lebesque measure. Thus $\mu_0^\Lambda(G^{\ell f}_{n,\Lambda}) = \frac{e^{-m(\Lambda)}}{n!}(m(\Lambda))^n$ and $\mu_0(\Omega^{\ell f}) = 1$. It is also clear that for

Λ_1, Λ_2 disjoint,

$$\mu_o^{\Lambda_1 \cup \Lambda_2}(d\omega) = \mu_o^{\Lambda_1}(d\omega_{\restriction \Lambda_1}) \mu_o^{\Lambda_2}(d\omega_{\restriction \Lambda_2}) .$$

We are really only interested in the μ_o for bounded Λ, but we could define the product measure $\mu_o = \prod_n \mu_o^{\Lambda_n}$ for a sequence of disjoint bounded Borel sets Λ_n covering $\mathbf{R}^\nu$ (the topology on $\Omega^{\ell f}$ is different from the product topology on $\prod_n \Omega^{\ell f}_{\Lambda_n}$, but the Borel σ-algebra will be the same). The set Ω, containing the configurations we are really interested in, would have μ_o-measure zero.

As in [21] or our Section II.2 we can define an *entropy* $S_\Lambda(\rho)$ for any probability measure ρ on $\Omega^{\ell f}$ and any bounded Borel subset Λ of $\mathbf{R}^\nu$ by

$$S_\Lambda(\rho) = - \int_{\Omega^{\ell f}_\Lambda} \rho^{(\Lambda)}(\omega) \log \rho^{(\Lambda)}(\omega) \mu_o^\Lambda(d\omega)$$

if the restriction of ρ to $\mathcal{S}_\Lambda$ is absolutely continuous with respect to μ_o^Λ, and $S_\Lambda(\rho) = -\infty$ otherwise. For any translation-invariant ρ, the *mean entropy*

$$s(\rho) = \lim_{\substack{\Lambda \to \infty \\ \text{(van Hove)}}} \frac{S_\Lambda(\rho)}{m(\Lambda)}$$

exists (in $[-\infty, 0]$), and has the same properties (with the same proofs) as in Section II.2.

An *interaction* Φ will be a real-valued translation-invariant function on the set $\mathfrak{F}$ of finite nonempty configurations in Ω. We will assume that Φ is continuous on each $\mathfrak{F}_n \equiv \{\omega \in \Omega : |\text{spt}(\omega)| = n\}$. The Banach space $\mathcal{B}$ will consist of those interactions Φ with

$$|||\Phi||| \equiv \sup_{\omega \in \Omega} \sum_{\substack{\omega' \subset \omega \\ 0 \in \text{spt}(\omega')}} \frac{|\Phi(\omega')|}{|\text{spt}(\omega')|} < \infty$$

while $\tilde{\mathcal{B}}$ will consist of those interactions with

$$\|\Phi\|_{\sim} \equiv \sup_{\omega\in\Omega} \sum_{\substack{\omega'\subset\omega \\ 0\in\mathrm{spt}(\omega')}} |\Phi(\omega')| < \infty .$$

The *Hamiltonian* for a bounded Borel subset Λ of $\mathbf{R}^{\nu}$ is the continuous function $H_{\Lambda}^{\Phi}(\omega) = \sum_{\substack{\omega'\subset\omega \\ \mathrm{spt}(\omega')\neq\emptyset}} \Phi(\omega')$ on Ω_{Λ}. Note that the sum converges absolutely, with

$$|H_{\Lambda}^{\Phi}(\omega)| \leq \sum_{\substack{\omega'\subset\omega \\ |\mathrm{spt}(\omega')|\neq 0}} \frac{1}{|\mathrm{spt}(\omega')|} \sum_{x\in\mathrm{spt}(\omega')} |\Phi(\omega')|$$

$$= \sum_{x\in\mathrm{spt}(\omega)} \sum_{\substack{\omega'\subset\omega \\ x\in\mathrm{spt}(\omega')}} \frac{|\Phi(\omega')|}{|\mathrm{spt}(\omega')|} \leq N_{\Lambda}|||\Phi|||$$

where $N_{\Lambda} = \sup\{|\mathrm{spt}(\omega)| : \omega\in\Omega_{\Lambda}\}$. If $a \equiv \inf_{Q\times Q\times S^{\nu-1}} f$ is the minimal allowed distance between particles, $N_{\Lambda} \leq m(B_{a/2})^{-1} m(\Lambda+B_{a/2})$ where $B_{a/2}$ is a ball of radius $a/2$. We defined H_{Λ}^{Φ} as above rather than as the function $\omega \mapsto H_{\Lambda}^{\Phi}(\omega_{\restriction\Lambda})$ on Ω, because the latter would not be continuous (the discontinuities can occur when particles cross into or out of Λ).

For interactions in $\tilde{\mathcal{B}}$, we can also define Hamiltonians with external-configuration or periodic boundary conditions. The Hamiltonian corresponding to the external configuration $\tau\in\Omega_{\Lambda^c}$ is

$${}_{\tau}H_{\Lambda}^{\Phi}(\omega) = \sum_{\substack{\omega'\subset\omega \\ \mathrm{spt}(\omega')\neq\emptyset}} \sum_{\substack{\tau'\subset\tau \\ |\mathrm{spt}(\tau')|<\infty}} \Phi(\omega'\times\tau')$$

defined only on $\{\omega \in \Omega_\Lambda : \omega \times \tau \in \Omega\}$. This sum converges absolutely to a continuous function, with

$$|{}_\tau H_\Lambda^\Phi(\omega)| \leq \sum_{x \in \mathrm{spt}(\omega)} \sum_{\substack{\omega' \subset \omega \cup \tau \text{ finite} \\ x \in \mathrm{spt}(\omega')}} |\Phi(\omega')| \leq N_\Lambda \|\Phi\|_\sim .$$

We could also define ${}_\tau H_\Lambda^\Phi(\omega) = +\infty$ if $\omega \in \Omega_\Lambda$ with $\omega \cup \tau \notin \Omega$. For periodic boundary conditions in a torus $T_n = \mathbf{R}^\nu / n\mathbf{Z}^\nu$, we first define the configuration space Ω_{T_n} (replacing $\mathbf{R}^\nu$ by T_n in the definition of Ω). There is a natural map $r_n : \Omega_{T_n} \to \Omega$ (repeating the configuration with period n in each direction). We define

$$H_{T_n}^\Phi(\omega) = \sum_{[\omega'] : \omega' \subset r_n(\omega)} \Phi(\omega')$$

where the sum is over equivalence classes $[\omega'] = \{\omega' + nj : j \in \mathbf{Z}^\nu\}$ of subconfigurations of $r_n(\omega)$. Again this converges absolutely to a continuous function, with $|H_{T_n}^\Phi(\omega)| \leq N_{T_n} \|\Phi\|_\sim$.

The *pressure* for our system in Λ is

$$P_\Lambda(\Phi) = m(\Lambda)^{-1} \ln \int_{\Omega_\Lambda} e^{-H_\Lambda^\Phi(\omega)} \mu_o^\Lambda(d\omega) .$$

Similarly, for $\Phi \in \tilde{\mathcal{B}}$ we have

$${}_\tau P_\Lambda(\Phi) = m(\Lambda)^{-1} \ln \int_{\{\omega \in \Omega_\Lambda : \omega \times \tau \in \Omega\}} e^{-{}_\tau H_\Lambda^\Phi(\omega)} \mu_o^\Lambda(d\omega)$$

and

$$P_{T_n}(\Phi) = n^{-\nu} \ln \int_{\Omega_{T_n}} e^{-H_{T_n}^\Phi(\omega)} \mu_o^{T_n}(d\omega)$$

where $\mu_0^{T_n}$ is the probability measure on $\Omega^{\ell f}_{T_n}$ defined analogously to μ_0^Λ.

The thermodynamic limit of the pressure can be obtained just as in Section I.2. Thus $P(\Phi) = \lim_{\Lambda \to \infty \text{ (van Hove)}} P_\Lambda(\Phi)$ exists and is a continuous convex function on $\mathcal{B}$. Our estimate on $|H^\Phi_\Lambda(\omega)|$ implies

$$|P(\Phi) - P(\Psi)| \leq m(B_{a/2})^{-1} |||\Phi - \Psi||| .$$

Better estimates on the close-packing density would improve the constant here. For an interaction in $\tilde{\mathcal{B}}$, any sequence of external configurations for regions tending (van Hove) to infinity, or periodic boundary conditions, yield the same limit $P(\Phi)$.

One more item is needed to formulate the variational principle, namely the average energy per unit volume ($\rho(A_\Phi)$ in the lattice system). This is not quite so easy to define in the continuous case. Let us fix a continuous non-negative function θ_0 on $\mathbf{R}^\nu$ with support contained in $B_{\frac{a}{2}}$ and $\int \theta_0(x)\,dx = 1$, and define $\theta(\omega) = \sum_{x \in \mathrm{spt}(\omega)} \theta_0(x)$. Thus θ is continuous on Ω, and by the hard core condition there is at most one nonzero term in this sum. Now we define

$$A_\Phi(\omega) = \sum_{\omega' \subset \omega} \frac{\theta(\omega')\Phi(\omega')}{|\mathrm{spt}(\omega')|} .$$

This is a continuous function on Ω for any $\Phi \in \mathcal{B}$, with $\|A_\Phi\|_\infty \leq \|\theta_0\|_\infty |||\Phi|||$. We claim that for any translation-invariant probability measure ρ on Ω, the energy per unit volume $m(\Lambda)^{-1} \int_\Omega H^\Phi_\Lambda(\omega|_\Lambda)\rho(d\omega)$ tends to $\rho(A_\Phi)$ as $\Lambda \to \infty$ (van Hove). In fact, let

$$F_\Lambda(\omega) = m(\Lambda)^{-1} \int_\Lambda dx\, A_\Phi(\omega - x) = \sum_{\omega' \subset \omega} \frac{\int_\Lambda dx\, \theta(\omega' - x)}{m(\Lambda)|\mathrm{spt}(\omega')|} \Phi(\omega') .$$

By Fubini's Theorem and translation invariance, $\rho(F_\Lambda) = \rho(A_\Phi)$. If $\mathrm{spt}(\omega') + B_{\frac{a}{2}} \subset \Lambda$ we have $\int_\Lambda dx\, \theta(\omega' - x) = |\mathrm{spt}(\omega')|$, so the term for ω' in the above sum also occurs in the sum for $m(\Lambda)^{-1} H_\Lambda^\Phi(\omega_{\restriction\Lambda})$. If $\mathrm{spt}(\omega') + B_{\frac{a}{2}}$ and Λ are disjoint, the term is zero. Thus in estimating $|F_\Lambda(\omega) - m(\Lambda)^{-1} H_\Lambda^\Phi(\omega_{\restriction\Lambda})|$ we need only consider those ω' containing a particle within $\frac{a}{2}$ of the boundary $\partial\Lambda$ of Λ. Now if $\Phi \in \tilde{\mathcal{B}}$

$$|F_\Lambda(\omega) - m(\Lambda)^{-1} H_\Lambda^\Phi(\omega_{\restriction\Lambda})| \leq m(\Lambda)^{-1} \sum_{\substack{x \in \mathrm{spt}(\omega) \\ \mathrm{dist}(x,\partial\Lambda) \leq \frac{a}{2}}} \sum_{\substack{\omega' \subset \omega \\ x \in \mathrm{spt}(\omega')}} |\Phi(\omega')|$$

$$\leq m(\Lambda)^{-1} N_{B_{\frac{a}{2}} + \partial\Lambda} \|\Phi\|_-$$

which tends to zero as $\Lambda \to \infty$ (van Hove). This proves the claim for $\Phi \in \tilde{\mathcal{B}}$, but the estimates $\|A_\Phi - A_\Psi\|_\infty \leq \|\theta_0\|_\infty |||\Phi - \Psi|||$ and $\|H_\Lambda^\Phi - H_\Lambda^\Psi\|_\infty \leq N_\Lambda |||\Phi - \Psi|||$ extend it to all $\Phi \in \mathcal{B}$.

An important tool in our proofs for the lattice systems was the ability, given any real-valued $B \in C(\Omega_\Lambda)$ for Λ finite, to form an interaction $\Psi = \Psi_B^\Lambda$ such that $\rho(A_\Psi) = \rho(B)$ for any invariant state ρ. This is not quite so easy or natural in the continuous case. Essentially the same difficulty occurs in the lattice system if we limit ourselves to "lattice-gas" interactions. We generalized our notion of interaction in the lattice system to allow "holes" as well as "occupied" sites to make contributions. There does not seem to be a very natural way of doing this for the continuous system. Instead, we will express our desired interaction in the "language" we have available. The only disadvantage is that the norm of the interaction may be large compared to that of the original function, so we will not obtain an isometry in the analogue of Lemma II.1.1, and the results of Section V.2 do not extend to the continuous system.

Let $\mathfrak{A}_\Lambda$ be the subalgebra of $C(\Omega)$ consisting of functions which only depend on the configuration in Λ. By the Stone-Weierstrass Theorem, the union of the $\mathfrak{A}_\Lambda$ for any sequence of bounded regions $\Lambda \to \infty$ (eventually containing each point of $\mathbf{R}^\nu$) is dense in $C(\Omega)$. Suppose $B \in \mathfrak{A}_\Lambda$ is real-valued with $B(\omega_\emptyset) = 0$ (recall $\omega_\emptyset$ is the empty configuration). We define an interaction Ψ_B by

$$\Psi_B(\omega) = \int_{\mathbf{R}^\nu} dx \sum_{\omega' \subset \omega} (-1)^{|\mathrm{spt}(\omega)| - |\mathrm{spt}(\omega')|} B(\omega' - x) .$$

Note that $B(\omega' - x) = 0$ if $\mathrm{spt}(\omega') - x$ is disjoint from Λ, so this is well-defined for any finite configuration ω, translation-invariant and continuous on each $\mathfrak{F}_n$. Moreover, cancellations occur in the sum so that $\Psi_B(\omega) = 0$ unless $\mathrm{spt}(\omega)$ is contained in some translate of Λ. Thus $\Psi_B \in \tilde{\mathfrak{B}}$. By the inclusion-exclusion principle we find that

$$H_{\Lambda'}^{\Psi_B}(\omega) = \sum_{\omega' \subset \omega} \Psi_B(\omega') = \int_{\mathbf{R}^\nu} dx\, B(\omega - x) \quad \text{for } \omega \in \Omega_{\Lambda'} .$$

Since $B(\omega_{\restriction \Lambda' - x}) = B(\omega)$ if $\Lambda \subset \Lambda' - x$, or 0 if $\Lambda \cup (\Lambda' - x) = \emptyset$, for any invariant state ρ we have

$$m(\Lambda')^{-1} \int_\Omega \rho(d\omega) H_{\Lambda'}^{\Psi_B}(\omega_{\restriction \Lambda'}) = m(\Lambda')^{-1} \int_{\mathbf{R}^\nu} dx \int_\Omega \rho(d\omega) B(\omega_{\restriction \Lambda' - x})$$

$$\to \rho(B) \quad \text{as } \Lambda' \to \infty \quad \text{(van Hove)} .$$

Thus $\rho(A_{\Psi_B}) = \rho(B)$.

With these definitions, we can use essentially the same proofs in dealing with the variational principle for hard-core continuous systems as for lattice systems, and obtain similar results (the main exception being Lemma II.1.1 and its consequences in Section V.2, as noted above). For example, in Theorem II.1.2 (showing that a P-bounded functional on $\mathcal{B}$ determines a unique invariant state) we would proceed as follows: given the P-bounded functional α, define ρ on real-valued functions in $\bigcup_{\substack{\Lambda \subset \mathbf{R}^\nu \\ \text{bounded}}} \mathfrak{A}_\Lambda$ by $\rho(A) = A(\omega_\emptyset) - \alpha(\Psi_B)$ where $B = A - A(\omega_\emptyset)$. Note $\rho(1) = 1$. If $A \geq 0$ we have

$$H_{\Lambda'}^{\Psi_B}(\omega) = \int_{\Lambda' - \Lambda} dx\, B(\omega - x) \geq -m(\Lambda' - \Lambda)\, A(\omega_\emptyset)$$

so $P(t\Psi_B) \leq \lim_{\Lambda' \to \infty (vH)} m(\Lambda')^{-1} m(\Lambda' - \Lambda)\, tA(\omega_\emptyset) = tA(\omega_\emptyset)$ for $t \geq 0$ and by P-boundedness of α we obtain $t\alpha(\Psi_B) \leq tA(\omega_\emptyset) + C$ and thus $\rho(A) \geq 0$. Thus ρ is a positive functional on a dense subalgebra of $C(\Omega)$ containing constants, so is continuous and extends by continuity to a state on $C(\Omega)$.

In formulating the DLR equations for our system, we require measures $\rho_\Lambda(\omega, d\tau)$ on Ω_{Λ^c} for $\omega \in \Omega_\Lambda$ and $\Lambda \subset \mathbf{R}^\nu$ open and bounded. These are to be supported on $\{\tau \in \Omega_{\Lambda^c} : \omega \cup \tau \in \Omega\}$ and chosen so that

$$\rho_\Lambda(\omega, d\tau) = e^{-{}_\tau H_\Lambda^\Phi(\omega)} \rho_\Lambda(\omega_\emptyset, d\tau)$$

(where ${}_\tau H_\Lambda^\Phi(\omega) = +\infty$ if $\omega \cup \tau \notin \Omega$). It can be shown (not quite as easily as for lattice systems) that $\omega \mapsto \rho_\Lambda(\omega, d\tau)$ is continuous from Ω_Λ to measures on Ω_{Λ^c} in the weak-* topology, so the choice is unique. Since there are no physically equivalent interactions, condition (i) in the analogue of Theorem III.4.1 for our system becomes (i′) $\Phi = \Psi$. The main point of the proof is then to determine the interaction Φ from the various Hamiltonians ${}_\tau H_\Lambda^\Phi$. In fact, we can use an empty external configuration and obtain Φ from H_Λ^Φ by the inclusion-exclusion principle:

$$\Phi(\omega) = \sum_{\omega' \subset \omega} (-1)^{|spt(\omega)| - |spt(\omega')|} H_\Lambda^\Phi(\omega)$$

where Λ is any bounded open set containing $\mathrm{spt}(\omega)$.

The most interesting application of the methods of Chapter V to our system is producing a "liquid-gas" type phase transition, with a long-range attractive two-body interaction, a one-body "chemical potential," and two invariant equilibrium states differing in density. For an invariant state ρ, the density is $\rho(\theta)$, and $f(x) = \rho(\theta\tau_x\theta)$ is a somewhat "smeared" two-point correlation function. The ergodic decomposition of ρ will include states of different densities if $\lim_{n\to\infty} \frac{1}{m(B_n)} \int_{B_n} f(x)\,dx > \rho(\theta)^2$. The methods of Chapter V lead to the following result:

THEOREM B.1. *Let* $\mathfrak{F}$ *be the cone of interactions* Ψ *of the form*

$\Psi(\omega) = \mu \geq 0$ *if* $|\mathrm{spt}(\omega)| = 1$

$\Psi(\omega) = 0$ *if* $|\mathrm{spt}(\omega)| > 2$

$\Psi(\omega) = V(x-y)$ *if* $\mathrm{spt}(\omega) = \{x,y\}$; *where* V *is a nonnegative continuous function on* R^ν, $V(x) = V(-x)$, *and*

$$|||\Psi||| = |\mu| + \frac{1}{2} \sup_{\substack{\omega\in\Omega \\ 0\in\mathrm{spt}(\omega)}} \sum_{\substack{x\in\mathrm{spt}(\omega)\\ x\neq 0}} |V(x)| < \infty .$$

Then for any $\Phi_0 \in \mathfrak{B}$ *there is* $\tilde{\Phi} \in \Phi_0 + \mathfrak{F}$ *which has at least two invariant equilibrium states with different density.*

Various other constraints may be imposed on V, such as spherical symmetry. We may obtain a breaking of translation invariance as in Theorem V.3.3, with an invariant equilibrium state ρ for which $\lim_{|x|\to\infty} f(x)$ does not exist, using a two-body interaction. However, this is not really a satisfactory sort of "crystallization": for that we would want a spontaneous breaking of rotational symmetry as well. To deal with crystallization in a satisfactory way would require a deep geometric and combinatorial analysis of the effect of the hard-core and short-range forces.

BIBLIOGRAPHY

[1] Anderson, R. D., Klee, V. L. Jr.: "Convex functions and upper semi-continuous collections," Duke Math. J. 19 (1952), 349-357.

[2] Araki, H.: "On the Equivalence of the KMS Condition and the Variational Principle for Quantum Lattice Systems," Commun. math. Phys. 38 (1974), 1-10.

[3] Araki, H., Miyata, H.: "On KMS boundary condition," Publ. RIMS Kyoto Univ. Ser. A 4 (1968), 373-385.

[4] Bishop, E., Phelps, R. R.: "The support functionals of a convex set," in: *AMS Proceedings of Symposia in Pure Mathematics*, vol. 7, pp. 27-35. Amer. Math. Soc., Providence, R. I., 1963.

[5] Dyson, F. J.: "Existence of a Phase Transition in a One-Dimensional Ising Ferromagnet," Commun. math. Phys. 12 (1969), 91-107.

[6] Federer, H.: *Geometric Measure Theory*. Berlin-Heidelberg-New York: Springer 1969.

[7] Fisher, M. E.: "On Discontinuity of the Pressure," Commun. math. Phys. 26 (1972), 6-14.

[8] Gallavotti, G., Miracle-Solé, S.: "Statistical Mechanics of Lattice Systems," Commun. math. Phys. 5 (1967), 317-323.

[9] ________________ : "A variational principle for the equilibrium of hard sphere systems," Ann. Inst. H. Poincaré Sect. A 8 (1968), 287-299.

[10] Greenberg, W.: "Critical Temperature Bounds of Quantum Lattice Gases," Commun. math. Phys. 13 (1969), 335-344.

[11] Griffiths, R. B.: "Correlations in Ising Ferromagnets III," Commun. math. Phys. 6 (1967), 121-127.

[12] Griffith, R. B., Ruelle, D.: "Strict Convexity ('Continuity') of the Pressure in Lattice Systems," Commun. math. Phys. 23 (1971), 169-175.

[13] Grothendieck, A.: "un resultat sur le dual d'une C*-algèbre," J. de Math. pures et appl. 36 (1957), 97-108.

[14] Hang, R., Hugenholtz, N. M., Winnink, M.: "On the Equilibrium States in Quantum Statistical Mechanics," Commun. math. Phys. 5 (1967), 215-236.

[15] Hurewicz, W., Wallman, H.: *Dimension Theory*. Princeton: Princeton University Press 1941.

[16] Lanford, O. E. III: "Entropy and Equilibrium States in Classical Statistical Mechanics," in: *Statistical Mechanics and Mathematical Problems* (1971 Battelle Rencontres), A. Lenard (Ed.), Lecture Notes in Physics vol. 20, pp. 1-113. Berlin-Heidelberg-New York: Springer 1973.

[17] Lanford, O. E. III, Robinson, D. W.: "Statistical Mechanics of Quantum Spin Systems III," Commun. math. Phys. 9 (1968), 327-338.

[18] Lanford, O. E. III, Ruelle, D.: "Observables at Infinity and States with Short Range Correlations in Statistical Mechanics," Commun. math. Phys. 13 (1969), 194-215.

[19] Mazur, S.: "Uber konvexe Mengen in linearen normierten Räumen," Studia Math. 4 (1933), 70-84.

[20] Miracle-Solé, S., Robinson, D. W.: "Statistical Mechanics of Quantum Mechanical Particles with Hard Cores II. The Equilibrium States," Commun. math. Phys. 19 (1970), 204-218.

[21] Robinson, D. W., Ruelle, D.: "Mean Entropy of States in Classical Statistical Mechanics," Commun. math. Phys. 5 (1967), 288-300.

[22] Rogers, C. A.: *Hausdorff Measures*. Cambridge: Cambridge Univ. Press 1970.

[23] Roos, H.: "Strict Convexity of the Pressure: A Note on a Paper of R. B. Griffiths and D. Ruelle," Commun. math. Phys. 36 (1974), 263-276.

[24] Ruelle, D.: "A Variational Formulation of Equilibrium Statistical Mechanics and the Gibbs Phase Rule," Commun. math. Phys. 5 (1967), 324-329.

[25] ________ : *Statistical Mechanics: Rigorous Results*. New York-Amsterdam: Benjamin 1969.

[26] Ruelle, D.: "Symmetry Breakdown in Statistical Mechanics," in: *Cargèse Lectures in Physics* vol. 4, D. Kastler (Ed.), pp. 169-194. New York-London-Paris: Gordon and Breach 1970.

[27] ________ : "Integral Representation of States on a C^*-Algebra," J. Funct. Anal. 6 (1970), 116-151.

[28] Simon, B.: Lectures at Princeton University, Spring 1974.

[29] Slawny, J.: Lectures at Princeton University, Fall 1973.

[30] Takesaki, M.: "States and Automorphisms of Operator Algebras, Standard Representations and the Kubo-Martin-Schwinger Boundary Condition," in: *Statistical Mechanics and Mathematical Problems* (1971 Battelle Rencontres), A. Lenard (Ed.), Lecture Notes in Physics vol. 20, pp. 205-246. Berlin-Heidelberg-New York: Springer 1973.

[31] Topping, D. M.: *Lectures on von Neumann Algebras.* London: Van Nostrand 1971.

[32] Winnink, M.: "Algebraic Aspects of the Kubo-Martin-Schwinger Condition," in: *Cargèse Lectures in Physics* vol. 4, D. Kastler (Ed.), pp. 235-255. New York-London-Paris: Gordon and Breach 1970.

[33] Yosida, K.: *Functional Analysis.* Berlin-Göttingen-Heidelberg: Springer 1965.

[34] Phelps, R. R.: *Lectures on Choquet's Theorem.* Princeton: Van Nostrand 1966.

[35] Lebowitz, J. L., Martin-Lof, A.: "On the Uniqueness of the Equilibrium State for Ising Spin Systems," Commun. math. Phys. 25 (1972), 276-282.

[36] Brascamp, H. J.: "Equilibrium States for a Classical Lattice Gas," Commun. math. Phys. 18 (1970), 82-96.

[37] Lieb, E. H., Ruskai, M. B.: "Proof of the strong subadditivity of quantum mechanical entropy," J. Math. Phys. 14 (1973), 1959-1964.

INDEX

Library of Congress Cataloging in Publication Data

Israel, Robert B
Convexity in the theory of lattice gases.

(Princeton series in physics)
Bibliography: p.
Includes index.
1. Lattice gas. 2. Convex domains. 3. Statistical mechanics. 4. Statistical thermodynamics. I. Title.
QC174.85.L38I87 530.1'3 78-51171
ISBN 0-691-08209-X
ISBN 0-691-08216-2 pbk.

Milton Keynes UK
Ingram Content Group UK Ltd.
UKHW021025140924
448309UK00006B/285

9 780691 606194